Teacher Edition

Eureka Math
Grade 2
Module 3

Special thanks go to the Gordon A. Cain Center and to the Department of Mathematics at Louisiana State University for their support in the development of *Eureka Math*.

For a free *Eureka Math* Teacher
Resource Pack, Parent Tip
Sheets, and more please
visit www.Eureka.tools

Published by the non-profit Great Minds

Copyright © 2015 Great Minds. No part of this work may be reproduced, sold, or commercialized, in whole or in part, without written permission from Great Minds. Non-commercial use is licensed pursuant to a Creative Commons Attribution-NonCommercial-ShareAlike 4.0 license; for more information, go to http://greatminds.net/maps/math/copyright. "Great Minds" and "Eureka Math" are registered trademarks of Great Minds.

Printed in the U.S.A.

This book may be purchased from the publisher at eureka-math.org

10 9

ISBN 978-1-63255-356-0

Eureka Math: A Story of Units Contributors

Katrina Abdussalaam, Curriculum Writer
Tiah Alphonso, Program Manager—Curriculum Production
Kelly Alsup, Lead Writer / Editor, Grade 4
Catriona Anderson, Program Manager—Implementation Support
Debbie Andorka-Aceves, Curriculum Writer
Eric Angel, Curriculum Writer
Leslie Arceneaux, Lead Writer / Editor, Grade 5
Kate McGill Austin, Lead Writer / Editor, Grades PreK–K
Adam Baker, Lead Writer / Editor, Grade 5
Scott Baldridge, Lead Mathematician and Lead Curriculum Writer
Beth Barnes, Curriculum Writer
Bonnie Bergstresser, Math Auditor
Bill Davidson, Fluency Specialist
Jill Diniz, Program Director
Nancy Diorio, Curriculum Writer
Nancy Doorey, Assessment Advisor
Lacy Endo-Peery, Lead Writer / Editor, Grades PreK–K
Ana Estela, Curriculum Writer
Lessa Faltermann, Math Auditor
Janice Fan, Curriculum Writer
Ellen Fort, Math Auditor
Peggy Golden, Curriculum Writer
Maria Gomes, Pre-Kindergarten Practitioner
Pam Goodner, Curriculum Writer
Greg Gorman, Curriculum Writer
Melanie Gutierrez, Curriculum Writer
Bob Hollister, Math Auditor
Kelley Isinger, Curriculum Writer
Nuhad Jamal, Curriculum Writer
Mary Jones, Lead Writer / Editor, Grade 4
Halle Kananak, Curriculum Writer
Susan Lee, Lead Writer / Editor, Grade 3
Jennifer Loftin, Program Manager—Professional Development
Soo Jin Lu, Curriculum Writer
Nell McAnelly, Project Director

This page intentionally left blank

Mathematics Curriculum

GRADE 2

Table of Contents

GRADE 2 • MODULE 3

Place Value, Counting, and Comparison of Numbers to 1,000

Grade 2 • Module 3

Place Value, Counting, and Comparison of Numbers to 1,000

OVERVIEW

In Module 2, students added and subtracted measurement units within 100 (**2.MD.5**, **2.MD.6**), a meaningful application of their work from Module 1 (**2.NBT.5**) and a powerful bridge to the base ten units of Grade 2.

In this 25-day Grade 2 module, students expand their skill with and understanding of units by bundling ones, tens, and hundreds up to a thousand with straws. Unlike the length of 10 centimeters in Module 2, these bundles are discrete sets. One unit can be grabbed and counted just like a banana—1 hundred, 2 hundred, 3 hundred, etc. (**2.NBT.1**). A number in Grade 1 generally consisted of two different units, tens and ones. Now, in Grade 2, a number generally consists of three units: hundreds, tens, and ones (**2.NBT.1**). The bundled units are organized by separating them largest to smallest, ordered from left to right. Over the course of the module, instruction moves from physical bundles that show the proportionality of the units to non-proportional place value disks and to numerals on the place value chart (**2.NBT.3**).

Furthermore, in this module instruction includes a great deal of counting: by ones, tens, and hundreds (**2.NBT.2**). Counting up using the centimeter tape or a classroom number line shows movement from left to right as the numbers increase. Counting up on the place value chart shows movement from right to left as the numbers increase. For example, as 10 ones are renamed as 1 ten, the larger unit is housed in the place directly to the left. The goal is for students to move back and forth fluidly between these two models, the number line and the place value chart, using them to either to rename units and compare numbers (**2.NBT.4**).

In this module, the place value story has advanced. Along with changing 10 ones for 1 ten, students now also change 10 tens for 1 hundred. This changing leads to the use of counting strategies to solve word problems (**2.OA.1**). In the next module, this change leads to mental math and the formal algorithms for addition and subtraction. Comparison extends into finding 100 more and 100 less, 10 more and 10 less, etc. Just as in Grade 1, *more* and *less* translate into formal addition and subtraction at the onset of Module 4 (**2.NBT.8**).

How is this module's learning foundational to later grades? Understanding 3 tens or 3 units of 10 leads to an understanding of 3 fours or 3 units or groups of four (Grade 3 OA standards), 3 fourths or 3 units of one-fourth (Grade 3 NF standards). Learning that 12 tens = 120 leads to an understanding of 12 tenths = 1.2, 4 thirds = 4/3 = 1 1/3, or even 4 threes = 12. Counting up and down by ones, tens, and hundreds with both the number line and place value chart is essential from Grade 3 forward for rounding and mental math (Grade 3 NBT standards) to meaningful understanding of all operations with base ten whole numbers (Grade 4 NBT standards) and to understanding place value's extension into decimal fractions and operations (Grade 5 NBT standards).

Notes on Pacing for Differentiation

If pacing is a challenge, consider the following modifications and omissions. Omit the Application Problem in Lesson 7 in order to give more time to practice the multiple segments in the Concept Development.

Reduce the Concept Development of Lesson 9 by omitting the empty number line. Instead, have students draw the bills used to count up from one amount to the next as was done in Lesson 3 but with bundles. If the empty number line is omitted in Lesson 9, then the component following the Problem Set of Lesson 13, "Estimating Numbers on the Empty Number Line," should also be omitted along with related questions from the Debrief and Problem 2 of the Exit Ticket. Consider using the empty number line as an extension.

Omit Lesson 10, and use it instead as an extension for early finishers or as a center activity during a different time of day (e.g., RTI time, economics, morning work, or problem of the week).

Reduce Lesson 11 by omitting the use of Dienes blocks in the Concept Development. Distribute bills instead. Omit the discussion about the difference between modeling with the blocks and the bills. Have students only model with bills and place value disks in the Problem Set.

Omit, or move to morning work, the Application Problems in Lessons 12 and 14 to allow more time for the Concept Developments. Consolidate Lessons 17 and 18, or perhaps use Lesson 18 as an activity for centers to allow students continued practice comparing numbers when represented in different forms.

Consider using Lesson 21 as either a center activity or morning work.

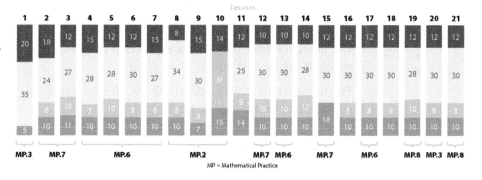

Distribution of Instructional Minutes

This diagram represents a suggested distribution of instructional minutes based on the emphasis of particular lesson components in different lessons throughout the module.

- Fluency Practice
- Concept Development
- Application Problems
- Student Debrief

MP = Mathematical Practice

©2015 Great Minds. eureka-math.org
G2-M3-TE-B2-1.3.1-01.2016

Focus Grade Level Standards

Understand place value.

2.NBT.1   Understand that the three digits of a three-digit number represent amounts of hundreds, tens, and ones; e.g., 706 equals 7 hundreds, 0 tens, and 6 ones. Understand the following as special cases:

  a.   100 can be thought of as a bundle of ten tens—called a "hundred."

  b.   The numbers 100, 200, 300, 400, 500, 600, 700, 800, 900 refer to one, two, three, four, five, six, seven, eight, or nine hundreds (and 0 tens and 0 ones).

2.NBT.2   Count within 1000; skip-count by 5s[1], 10s, and 100s.

2.NBT.3   Read and write numbers to 1000 using base-ten numerals, number names, and expanded form.

2.NBT.4   Compare two three-digit numbers based on meanings of the hundreds, tens, and ones digits, using >, =, and < symbols to record the results of comparisons.

Foundational Standards

1.NBT.2   Understand that the two digits of a two-digit number represent amounts of tens and ones. Understand the following as special cases:

  a.   10 can be thought of as a bundle of ten ones—called a "ten."

  b.   The numbers from 11 to 19 are composed of a ten and one, two, three, four, five, six, seven, eight, or nine ones.

  c.   The numbers 10, 20, 30, 40, 50, 60, 70, 80, 90 refer to one, two, three, four, five, six, seven, eight, or nine tens (and 0 ones).

1.NBT.3   Compare two two-digit numbers based on meanings of the tens and ones digits, recording the results of comparisons with the symbols >, =, and <.

Focus Standards for Mathematical Practice

MP.2   **Reason abstractly and quantitatively.** Mathematically proficient students make sense of quantities and their relationships in problem situations. They bring two complementary abilities to bear on problems involving quantitative relationships: the ability to *decontextualize*—to abstract a given situation and represent it symbolically and manipulate the representing symbols as if they have a life of their own, without necessarily attending to their referents—and the ability to *contextualize*—to pause as needed during the manipulation process in order to probe into the referents for the symbols involved. Quantitative reasoning entails habits of creating a coherent representation of the problem at hand; considering the 6 units involved; attending to the meaning of quantities, not just how to compute them; and knowing and flexibly using different properties of operations and objects (exemplified in Topic D).

---

[1] Use an analog clock to provide a context for skip-counting by fives.

MP.3   **Construct viable arguments and critique the reasoning of others.** Mathematically proficient students understand and use stated assumptions, definitions, and previously established results in constructing arguments. They make conjectures and build a logical progression of statements to explore the truth of their conjectures. They are able to analyze situations by breaking them into cases and can recognize and use counterexamples. They justify their conclusions, communicate them to others, and respond to the arguments of others. They reason inductively about data, making plausible arguments that take into account the context from which the data arose. Mathematically proficient students are also able to compare the effectiveness of two plausible arguments, distinguish correct logic or reasoning from that which is flawed, and—if there is a flaw in an argument—explain what it is. Elementary students can construct arguments using concrete referents such as objects, drawings, diagrams, and actions. Such arguments can make sense and be correct, even though they are not generalized or made formal until later grades. Later, students learn to determine domains to which an argument applies. Students at all grades can listen or read the arguments of others, decide whether they make sense, and ask useful questions to clarify or improve the argument (exemplified in Topics A and E).

MP.6   **Attend to precision.** Mathematically proficient students try to communicate precisely to others. They try to use clear definitions in discussion with others and in their own reasoning. They state the meaning of the symbols they choose, including using the equal sign consistently and appropriately. They are careful about specifying units of measure and labeling axes to clarify the correspondence with quantities in a problem. They calculate accurately and efficiently and express numerical answers with a degree of precision appropriate for the problem context. In the elementary grades, students give carefully formulated explanations to each other. By the time they reach high school, they have learned to examine claims and make explicit use of definitions (exemplified in Topics C and F).

MP.7   **Look for and make use of structure.** Mathematically proficient students look closely to discern a pattern or structure. Young students, for example, might notice that three and seven more is the same amount as seven and three more, or they may sort a collection of shapes according to how many sides the shapes have. Later, students will see 7×8 equals the well remembered $7 \times 5 + 7 \times 3$, in preparation for learning about the distributive property. In the expression $x^2 + 9x + 14$, older students can see the 14 as 2×7 and the 9 as $2 + 7$. They recognize the significance of an existing line in a geometric figure and can use the strategy of drawing an auxiliary line for solving problems. They also can step back for an overview and shift perspective. They can see complicated things, such as some algebraic expressions, as single objects or as being composed of several objects. For example, they can see $5 - 3(x - y)^2$ as 5 minus a positive number times a square and use that to realize that its value cannot be more than 5 for any real numbers x and y (exemplified in Topic B).

MP.8   **Look for and express regularity in repeated reasoning.** Mathematically proficient students notice if calculations are repeated and look both for general methods and for shortcuts. Upper elementary students might notice when dividing 25 by 11 that they are repeating the same calculations over and over again and conclude they have a repeating decimal. By paying attention to the calculation of slope as they repeatedly check whether points are on the line through (1, 2) with slope 3, middle school students might abstract the equation $(y - 2)/(x - 1) = 3$. Noticing the regularity in the way terms cancel when expanding $(x - 1)$

©2015 Great Minds. eureka-math.org
G2-M3-TE-B2-1.3.1-01.2016

$(x + 1)$, $(x - 1)$ $(x^2 + x + 1)$, and $(x - 1)$ $(x^3 + x^2 + x + 1)$ might lead them to the general formula for the sum of a geometric series. As they work to solve a problem, mathematically proficient students maintain oversight of the process, while attending to the details. They continually evaluate the reasonableness of their intermediate results (exemplified in Topic G).

Overview of Module Topics and Lesson Objectives

Standards		Topics and Objectives	Days
2.NBT.1	A	**Forming Base Ten Units of Ten, a Hundred, and a Thousand**	1
		Lesson 1: Bundle and count ones, tens, and hundreds to 1,000.	
2.NBT.2[2] 2.NBT.1	B	**Understanding Place Value Units of One, Ten, and a Hundred**	2
		Lesson 2: Count up and down between 100 and 220 using ones and tens.	
		Lesson 3: Count up and down between 90 and 1,000 using ones, tens, and hundreds.	
2.NBT.3 2.NBT.1 2.NBT.2	C	**Three-Digit Numbers in Unit, Standard, Expanded, and Word Forms**	4
		Lesson 4: Count up to 1,000 on the place value chart.	
		Lesson 5: Write base ten three-digit numbers in unit form; show the value of each digit.	
		Lesson 6: Write base ten numbers in expanded form.	
		Lesson 7: Write, read, and relate base ten numbers in all forms.	
2.NBT.2 2.NBT.1 2.NBT.3 2.MD.8	D	**Modeling Base Ten Numbers Within 1,000 with Money**	3
		Lesson 8: Count the total value of $1, $10, and $100 bills up to $1,000.	
		Lesson 9: Count from $10 to $1,000 on the place value chart and the empty number line.	
		Lesson 10: Explore $1,000. How many $10 bills can we change for a thousand dollar bill?	
		Mid-Module Assessment: Topics A–D (assessment ½ day, return ½ day, remediation or further applications 1 day)	2

---

[2] Use analog clock to provide a context for skip-counting by fives.

EUREKA MATH™

Standards			Topics and Objectives	Days
2.NBT.A	E		**Modeling Numbers Within 1,000 with Place Value Disks**	5
		Lesson 11:	Count the total value of ones, tens, and hundreds with place value disks.	
		Lesson 12:	Change 10 ones for 1 ten, 10 tens for 1 hundred, and 10 hundreds for 1 thousand.	
		Lesson 13:	Read and write numbers within 1,000 after modeling with place value disks.	
		Lesson 14:	Model numbers with more than 9 ones or 9 tens; write in expanded, unit, standard, and word forms.	
		Lesson 15:	Explore a situation with more than 9 groups of ten.	
2.NBT.4	F		**Comparing Two Three-Digit Numbers**	3
		Lesson 16:	Compare two three-digit numbers using <, >, and =.	
		Lesson 17:	Compare two three-digit numbers using <, >, and = when there are more than 9 ones or 9 tens.	
		Lesson 18:	Order numbers in different forms. (Optional)	
2.NBT.2 **2.OA.1** **2.NBT.8**	G		**Finding 1, 10, and 100 More or Less than a Number**	3
		Lesson 19:	Model and use language to tell about 1 more and 1 less, 10 more and 10 less, and 100 more and 100 less.	
		Lesson 20:	Model 1 more and 1 less, 10 more and 10 less, and 100 more and 100 less when changing the hundreds place.	
		Lesson 21:	Complete a pattern counting up and down.	
			End-of-Module Assessment: Topics A–G (assessment ½ day, return ½ day, remediation or further applications 1 day)	2
Total Number of Instructional Days				25

Terminology

New or Recently Introduced Terms

- Base ten numerals (e.g., a thousand is 10 hundreds, a hundred is 10 tens, starting in Grade 3 a one is 10 tenths, etc.)
- Expanded form (e.g., 500 + 70 + 6)
- Hundreds place (e.g., the 5 in 576 is in the hundreds place)
- One thousand (1,000)
- Place value or number disk (pictured)
- Standard form (e.g., 576)
- Unit form (e.g., 5 hundreds 7 tens 6 ones)
- Word form (e.g., five hundred seventy-six)

Unit form modeled with place value disks:
7 hundreds 2 tens 6 ones = 72 tens 6 ones

Familiar Terms and Symbols[3]

- =, <, > (equal, less than, greater than)
- Altogether (e.g., 59 centimeters and 17 centimeters; altogether there are 76 centimeters)
- Bundling, grouping (putting smaller units together to make a larger one, e.g., putting 10 ones together to make a ten or 10 tens together to make a hundred)
- How many more/less (the difference between quantities)
- How much more/less (the difference between quantities)
- More than/less than (e.g., 576 is more than 76; 76 is less than 576)
- Number sentence (an equation or inequality that has a true or false value and contains no unknowns, e.g., 3 + 2 = 5)
- Ones place (e.g., the 6 in 576 is in the ones place)
- Place value (the unitary values of the digits in numbers)
- Renaming, changing (instead of *carrying* or *borrowing*, e.g., a group of 10 ones is renamed a ten when the ones are bundled and moved from the ones to the tens place; if using $1 bills, they may be changed for a $10 bill when there are enough)
- Tens place (e.g., the 7 in 576 is in the tens place)
- Units of ones, tens, hundreds, one thousand (a single one and groups of 10s, 100s, and 1,000)

---

[3]These are terms and symbols students have seen previously.

Suggested Tools and Representations

- 2 boxes of 1,000 straws per class of 25
- Clock number line (details in Lesson 1 Fluency Practice)
- Dice, 1 per pair
- Dienes blocks
- Hide Zero cards (also known as place value cards) showing numbers 1–5, 10–50, and 100—500 (1 small set per student) (Lesson 4 Template 1))
- Hundreds place value chart (Lesson 4 Template 2)
- Meter strip (Lesson 1 Template)
- Number spelling activity sheet (Lesson 7 Activity Sheet)
- Personal white boards
- Place value box (details in Lesson 4 Concept Development)
- Place value cards to 1,000, 1 large teacher set
- Place value disks: suggested minimum of one set per pair (18 ones, 18 tens, 18 hundreds, and 1 one thousand)
- Play money: $1, $5, $10, and $100 bills (10 ones, 1 five, 12 tens, and 10 hundreds per pair), and a single set of 16 pennies, 13 dimes
- Rubber bands, 16 per pair
- Small plastic bags (small resealable bags)

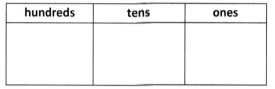

hundreds	tens	ones

Hundreds Place Value Chart

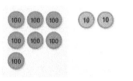

Place Value Disks

Unlabeled Hundreds Place Value Chart
(use with number disks)

Scaffolds[4]

The scaffolds integrated into *A Story of Units* give alternatives for how students access information as well as express and demonstrate their learning. Strategically placed margin notes are provided within each lesson elaborating on the use of specific scaffolds at applicable times. They address many needs presented by English language learners, students with disabilities, students performing above grade level, and students performing below grade level. Many of the suggestions are organized by Universal Design for Learning (UDL) principles and are applicable to more than one population. To read more about the approach to differentiated instruction in *A Story of Units,* please refer to "How to Implement *A Story of Units.*"

---

[4]Students with disabilities may require Braille, large print, audio, or special digital files. Please visit the website www.p12.nysed.gov/specialed/aim for specific information on how to obtain student materials that satisfy the National Instructional Materials Accessibility Standard (NIMAS) format.

Assessment Summary

Type	Administered	Format	Standards Addressed
Mid-Module Assessment Task	After Topic D	Constructed response with rubric	2.NBT.1 2.NBT.2 2.NBT.3
End-of-Module Assessment Task	After Topic G	Constructed response with rubric	2.NBT.1 2.NBT.2 2.NBT.3 2.NBT.4

EUREKA MATH

2
GRADE

Mathematics Curriculum

Topic A

Forming Base Ten Units of Ten, a Hundred, and a Thousand

2.NBT.1

Focus Standard:	2.NBT.1	Understand that the three digits of a three-digit number represent amounts of hundreds, tens, and ones; e.g., 706 equals 7 hundreds, 0 tens, and 6 ones. Understand the following as special cases:
		a. 100 can be thought of as a bundle of ten tens—called a "hundred."
		b. The numbers 100, 200, 300, 400, 500, 600, 700, 800, 900 refer to one, two, three, four, five, six, seven, eight, or nine hundreds (and 0 tens and 0 ones).
Instructional Days:	1	
Coherence -Links from:	G1–M6	Place Value, Comparison, Addition and Subtraction to 100
-Links to:	G2–M4	Addition and Subtraction Within 200 with Word Problems to 100

When students gather on the carpet in a circle, the teacher pours out a box of 1,000 straws. "How can we count these easily?" Students are led to suggest that bundles of 10 would make it much easier to count and recount the giant pile of straws. Students skip-count and experience that 1 hundred is equal to both 100 ones and 10 tens (**2.NBT.1a**). Likewise, 1 thousand is equal to both 100 tens and 10 hundreds (**2.NBT.1b**). Just as students added and subtracted centimeter units in Module 2, in Module 3 they skip-count using bundles of straws as units. The efficiency of place value and base ten numbers comes to life as students repeatedly bundle 10 ones to make 1 ten and subsequently bundle 10 tens to make 1 hundred.

A Teaching Sequence Toward Mastery of Forming Base Ten Units of Ten, a Hundred, and a Thousand
Objective 1: Bundle and count ones, tens, and hundreds to 1,000. (Lesson 1)

Topic A: Forming Base Ten Units of Ten, a Hundred, and a Thousand

11

Lesson 1

Objective: Bundle and count ones, tens, and hundreds to 1,000.

Suggested Lesson Structure

- ■ Fluency Practice   (20 minutes)
- ☐ Concept Development   (35 minutes)
- ▨ Student Debrief   (5 minutes)
- **Total Time**   **(60 minutes)**

Fluency Practice (20 minutes)

- Meter Strip Subtraction: Taking Multiples of 10 from Numbers Within 10 to 100 **2.NBT.5**   (5 minutes)
- Skip-Count Up and Down by Fives on the Clock **2.NBT.2**   (11 minutes)
- Happy Counting: Up and Down by Ones from 95 to 121 **2.NBT**   (2 minutes)
- Skip-Count by Tens: Up and Down Crossing 100 **2.NBT.2**   (2 minutes)

Meter Strip Subtraction: Taking Multiples of 10 from Numbers Within 10 to 100 (5 minutes)

Materials: (S) Meter strip (Fluency Template)

T: Put your finger on 0 to start. I'll say the whole measurement. Slide up to that number. Then take away 10 centimeters and tell me how many centimeters your finger is from 0.

T: Let's try one. Fingers at 0 centimeters! (Pause.) 50 centimeters.

S: (Slide their fingers to 50.)

T: Remember to take 10. (Pause.) How far is your finger from 0?

S: 40.

NOTES ON MULTIPLE MEANS OF ACTION AND EXPRESSION:

The pace in meter strip subtraction may be too rapid for some groups of students. If necessary, adjust it by providing more practice with multiples of 10 before moving on to other numbers.

Students start with their fingers at 0 and slide to the whole amount each time. This step maintains their knowledge of the distance between 0 and a given measurement. It provides visual and kinesthetic reinforcement of number sequence and relationships between numbers on the number line for students who may need it.

T: 40 what?

S: 40 centimeters!

T: Slide your finger back to 0. (Pause.) 85 centimeters.

T: (Pause.) How far is your finger from 0?

S: 75 centimeters!

T: Good. Slide back to 0. (Pause.) 49 centimeters.

Continue with examples as necessary.

T: Nice work. This time I'll say the whole measurement, and you take 20 centimeters. Ready?

T: Slide back to 0. (Pause.) 65 centimeters.

S: 45 centimeters!

Continue with the following possible sequence: Slide from 0 to 32, and then take 20; to 36 and then take 30; to 78 and then take 50; to 93 and then take 40; and to 67 and then take 60.

Skip-Count Up and Down by Fives on the Clock (11 minutes)

Materials: (T) A "clock" made from a 24-inch ribbon marked off at every 2 inches

T: (Display the ribbon as a horizontal number line—example pictured above.) Count by fives as I touch each mark on the ribbon.

S: (Starting with 0, count by fives to 60.)

T: (Make the ribbon into a circle resembling a clock.) Now I've shaped my ribbon to look like a …

S: Circle! Clock!

T: Let's call it a clock. Again, count by fives as I touch each mark on the clock.

S: (Starting with 0, skip-count by fives to 60.)

T: This time, the direction my finger moves on the clock will show you whether to count up or down. (While explaining, demonstrate sliding a finger forward and backward around the clock.)

T: As I slide to the marks, you count them by fives.

NOTES ON MULTIPLE MEANS OF REPRESENTATION:

Students have just finished working with meter strips, which are concrete number lines. In this activity, they move to working with an abstract number line: the clock. A clock is a circular number line. Visually demonstrate this for students by making the clock from a 24-inch ribbon marked off every 2 inches, similar to the one pictured with this activity.

Consider measuring the intervals in advance, making the marks very lightly so that they are hard for others to see. Then, begin the activity by making the marks dark enough for all to see as students count along by ones to notice that there are 12 marks.

Lesson 1: Bundle and count ones, tens, and hundreds to 1,000.

13

©2015 Great Minds. eureka-math.org
G2-M3-TE-B2-1.3.1-01.2016

Starting at 12, slide forward to 4 as students count on. On a clock, 12 represents both 0 and 60. We are not stating 0 so that students count on effectively.

S: 5, 10, 15, 20.

T: How many minutes is that?

S: 20 minutes!

T: (Starting from 4, slide a finger forward to 9. Do not restate 20. Count on.)

S: 25, 30, 35, 40, 45.

T: How many minutes is that?

S: 45 minutes!

T: (Keep a finger at 9.) What if I slide back one mark, then how many minutes?

S: 40 minutes!

T: Good. What if I slide forward one mark, then how many minutes?

S: 45 minutes!

T: Nice job. Let's count back from 50. (Start from 50 and slide back 5 times.)

S: 45, 40, 35, 30, 25.

T: How many minutes now?

S: 25 minutes!

Continue. Notice which switches or numbers students find most difficult, and use their cues to guide the practice provided.

T: Let's pause for a couple of minutes to think about the tools we've used so far today.

T: With your partner, compare the meter strip to the clock. How are they the same? How are they different?

MP.3

For about one or two minutes, circulate and listen for responses. Use questioning strategies to support student communication and the level of their insights.

S: They're both curly. Remember our paper meter strips were curly, too. → They can both be a straight line. → The clock has 12 marks and the other one has a lot more. → You can count with both of them. → The clock goes to 60 and the meter strip goes to 100. → On one you skip-count by fives and on the other you can skip-count by twos or tens. → All the marks on the clock are the same space apart, and the marks on the meter strip are the same space apart. → You can use them both to measure. → One measures time and one measures length.

NOTES ON
MULTIPLE MEANS
OF REPRESENTATION:

Partner Talk
Partner talk provides an opportunity for English language learners to rehearse language in a smaller, safer setting. It also provides an opportunity to pair children who can support one another with a shared first language. Balance pairings so that students feel supported but also benefit from the peer modeling and individualized practice with English provided by structured partner talk.
Partner talk serves struggling and advanced students by allowing them to work at their own levels. It's wise to consider students' strengths when assigning who will talk first. It can work well for Partner A to model strong language when partnered with English language learners or less verbally advanced students.

Questioning
If students have difficulty growing ideas or sustaining conversation, consider asking an advancing question: "Yes, you can count on both of them. What do you measure with each?"
This scaffold is especially relevant for students who have difficulty staying focused and students working below grade level. It also provides scaffolding for English language learners who, in order to respond, may rely on the vocabulary used in the question that is asked.

Lesson 1:   Bundle and count ones, tens, and hundreds to 1,000.

T: I hear some of you saying that we use both tools to measure. It's true that clocks and meter strips both measure.

T: What makes them useful for measuring? Talk with your partner for 30 seconds.

S: They both have marks that are the same space apart. → The numbers go from smallest to biggest. → They're both like rulers, but they have different units. → Clocks measure time. We can't see that! → It's like they both keep track of our counts. → And they both give us a place to count.

MP.3

T: I used a ribbon to make our clock. What would happen if I moved it back into a horizontal line so that it looked more like a meter strip? Partner A, could I still use it to measure the length of time? Tell Partner B why or why not.

S: I think so. You're not changing the numbers on it. You can still count how many minutes. When you've counted the whole thing, you know an hour went by.

T: (Move the ribbon back into a horizontal line and present it to students near the meter strip for a visual comparison.) Partner B, tell Partner A why you agree or disagree.

S: I disagree. There are no little hands to tell you where to count and tell you how many minutes have gone by.

T: Keep thinking and talking about these two measurement tools. Ask your parents what they think!

Happy Counting: Up and Down by Ones from 95 to 121 (2 minutes)

T: Let's count by ones, starting at 95. Ready? (Rhythmically point up until a change is desired. Show a closed hand and then point down. Continue, mixing it up.)

S: 95, 96, 97, 98, 99, 100, 101, 102. (Switch direction.) 101, 100. (Switch direction.) 101, 102, 103, 104, 105, 106, 107, 108, 109, 110, 111, 112. (Switch direction.) 111, 110, 109. (Switch direction.) 110, 111, 112, 113, 114, 115, 116, 117. (Switch direction.) 116, 115, 114. (Switch direction.) 115, 116, 117, 118, 119, 120, 121. (Switch direction.) 120, 119, 118.

NOTES ON MULTIPLE MEANS OF ENGAGEMENT:

"It's true that we use both tools to measure. It's true that clocks and meter strips both measure lengths."

This is an example of telling rather than eliciting, a unifying way to follow partner sharing. The telling makes a certain fact common knowledge from which new ideas grow. It's okay to tell rather than elicit. Strategically telling is a facilitation technique that keeps the conversation moving. Use it to correct misconceptions and set students up to go deeper along a line of reasoning.

NOTES ON MULTIPLE MEANS OF REPRESENTATION:

Move the ribbon back into a horizontal line and present it to students near the meter strip for a visual comparison.

Providing a visual representation allows English language learners to access the content while learning important vocabulary. In this case, a visual comparison also helps clarify the topic of discussion. Maximize the benefits of visual comparison by placing the ribbon alongside the meter strip before Partner A shares with Partner B.

Skip-Count by Tens: Up and Down Crossing 100
(2 minutes)

T: Let's skip-count by tens starting at 60.

T: Ready? (Rhythmically point up until a change is desired. Show a closed hand and then point down. Continue, mixing it up.)

S: 60, 70, 80, 90, 100, 110, 120, 130, 140. (Switch direction.) 130, 120, 110, 100, 90. (Switch direction.) 100, 110, 120, 130, 140, 150, 160, 170, 180, 190, 200, 210, 220. (Switch direction.) 210, 200, 190, 180.

Concept Development (35 minutes)

Materials: (T) Box of 1,000 straws or sticks

Students are seated in a U shape or circle on the carpet. Quite dramatically empty the contents of the box onto the carpet.

T: Let's count these straws! About how many do you think there might be? Discuss your ideas with your partner.

T: Let's see how many there really are.

T: How can we count them in a way that is fast and accurate, or efficient, so that we can get to recess on time?

S: We could split them up into piles and share the work. → By twos! → By fives! → By tens. → By ones.

T: There are some very clear ideas. Discuss with your partner which method would be the most efficient, counting by ones, twos, fives, or tens.

T: I hear most groups agreeing that counting by tens is the most efficient. Why is it more efficient to count by units of ten than units of two?

S: Because there will be more units of two, it will take longer. → The tens are the biggest so there are fewer of them to confuse us when we count.

T: Are you ready to get going? Let's count 10 straws and then wrap them in a rubber band to make a new unit of ten. I will put a pile of straws and rubber bands in front of each group of 3 students.

S: (Work for about 8 minutes to finish bundling all the straws.)

T: Let's make even larger units: Hundreds. It takes 10 tens to make a hundred. Count with me.

NOTES ON
MULTIPLE MEANS
OF ENGAGEMENT:

As often as possible, create opportunities for every student to respond every time. The vignettes throughout the entire module facilitate this by continuously demonstrating varied response patterns and materials including choral response, partner talk, personal white boards, and individual tools like meter strips. Response patterns built on 100% student participation have powerful effects on student engagement and lesson pacing.

Choral response allows English language learners to listen to correct pronunciation and language structure while practicing with the support of peer voices. Choral response that incorporates chanting, like the counting activities presented to the left, allows struggling students and those with auditory processing difficulty to be supported by the group as they pick up on language and patterns.

Wait time is an important component of choral response. It provides children with an opportunity to independently process the question and formulate an answer before speaking. This is a useful scaffold for English language learners and struggling students. Wait time is built into many vignettes where the dialogue says, "Pause," or when the teacher asks students to wait for a signal to respond.

EUREKA
MATH™

S: (Place a ten before each count.) 1 ten, 2 tens, 3 tens, 4 tens, 5 tens, 6 tens, 7 tens, 8 tens, 9 tens, 10 tens.

T: What is the value of 10 tens?

S: 1 hundred.

T: How many straws equal 1 ten?

S: 10 straws.

T: Now, let's count the number of straws in 10 tens or 1 hundred.

S: (Repeat the process.) 10, 20, 30, 40, 50, 60, 70, 80, 90, 100.

T: So, how many straws are in 10 tens?

S: 100 straws.

T: What is another way to say 10 tens?

S: 1 hundred.

T: As a group, bundle 10 tens to make 1 hundred. Put the tens and ones you have left over to one side.

S: (Work.)

T: Tell your neighboring group how many of each unit—ones, tens, and hundreds—you have. The single straws are units of one.

S: We have 1 hundred, 6 tens, and 4 ones.

T: Let's make the single straws into as many tens as we can. How many extra ones does your group have?

S: 3.

T: Students, what do we need to add to 3 ones to make 10 ones? (Pause.)

S: 7 ones.

T: Which group has 7 ones? (Or, can we combine 2 groups' straws to get 7 ones?)
     Pass them to Group 1.

Repeat the *make ten* process with all the extra ones.

T: Now that we have made as many units of ten as possible, let's make more units of one hundred.

T: Group 2, how many tens do you have that are not bundled as 1 hundred?

S: 6 tens.

T: Students, at the signal, what do we need to add to 6 tens to make 10 tens? (Signal.)

S: 4 tens.

T: 6 tens plus 4 tens is?

S: 10 tens.

T: What is another way to say 10 tens?

S: 1 hundred.

T: How can you prove that 10 tens is the same as 100?

S: I could unbundle the hundred and count all the tens. → I can skip-count by 10 and count how many times it takes to get to 100. → When I skip-count on my fingers it takes all 10 to get to 100.

**NOTES ON
MULTIPLE MEANS
OF ENGAGEMENT:**

All through this module, students must pay attention to the units they are counting and use precise language to convey their knowledge. Hold them accountable: 6 tens + 4 tens is 10 tens.

Repeat the *make 1 hundred* process, bundling all the tens as hundreds.

T: Now that we have made as many hundreds as possible, let's make units of **one thousand**.

T: Think about the structure and pattern of numbers as we've moved from ones to tens to hundreds. Then talk with your partner: How many hundreds do you think make 1 thousand? Be ready to explain why.

S: When we count, the numbers always go 1, 2, 3, 4, 5, 6, 7, 8, 9, 10 and then we get a new unit. → There are 10 hundreds in 1 thousand because we always make one bigger group out of 10 smaller groups.

T: Yes, 10 of a smaller unit make 1 of the next largest unit. I like the way you used what you've learned about the structure of numbers to figure out something new.

T: So, how many hundreds are in 1 thousand? Give me a complete sentence.

S: 10 hundreds are in 1 thousand!

T: Group 3, how many hundreds do you have?

S: 2 hundreds.

T: Students, complete the sentence: 2 hundreds plus how many hundreds equals 10 hundreds? (Pause.)

S: 2 hundreds plus 8 hundreds equals 10 hundreds.

T: Hand all your hundreds over! (Bundle them up to make one thousand.)

T: Count the hundreds for me. I'll listen.

S: 1 hundred, 2 hundreds, ...

T: How many hundreds do we have here?

S: 10 hundreds!

T: Another name for 10 hundreds is 1 thousand, a new unit!

T: At the signal, what is the largest unit we worked with today? (Signal.)

S: 1 thousand!

T: The next largest?

S: 1 hundred!

T: The next?

S: 1 ten!

T: The smallest?

S: 1 one!

T: (Give each pair 1 straw, a bundle of 1 ten, and a bundle of 1 hundred.) Show and tell your partner our units in order from smallest to largest and largest to smallest.

T: How many different units did we work with today?

S: 4 units!

T: Tell me the unit names from smallest to largest.

S: Ones, tens, hundreds, and thousands.

©2015 Great Minds. eureka-math.org
G2-M3-TE-B2-1.3.1-01.2016

Problem Set (10 minutes)

Students should do their personal best to complete the Problem Set within the allotted 10 minutes. Some problems do not specify a method for solving. This is an intentional reduction of scaffolding that invokes MP.5, Use Appropriate Tools Strategically. Students should solve these problems using the RDW approach used for Application Problems.

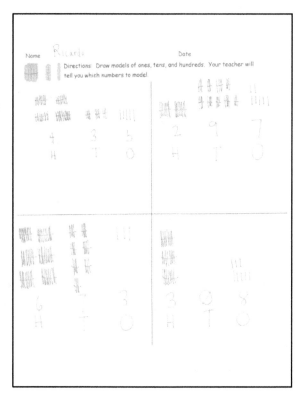

For some classes, it may be appropriate to modify the assignment by specifying which problems students should work on first. With this option, let the purposeful sequencing of the Problem Set guide your selections so that problems continue to be scaffolded. Balance word problems with other problem types to ensure a range of practice. Consider assigning incomplete problems for homework or at another time during the day.

- T: We used straws to show units of hundreds, tens, and ones. Now, let's draw models of these units.
- T: (Draw a sample of each unit, as shown in the picture.)
- T: Draw and label 4 hundreds. Whisper count as you draw.
- S: (Whisper count and draw as you model.)
- T: Whisper count, draw, and label 3 tens.
- S: (Whisper count and draw.)
- T: Now, whisper count, draw, and label 5 ones. If you don't have enough room in the box, use your eraser and try again.
- S: (Whisper count and draw.)
- T: Tell me the number of each unit in order from largest to smallest.
- S: 4 hundreds, 3 tens, 5 ones.
- T: The name of that number is?
- S: 435.
- T: Yes.
- T: In the next box down, draw and label 6 hundreds, 7 tens, 3 ones.
- S: (Work.)
- T: When I say, "Show me your work," hold up your paper so I can see your independent effort.

Repeat the process with the following: 297 and 308.

Lesson 1:   Bundle and count ones, tens, and hundreds to 1,000.

19

©2015 Great Minds. eureka-math.org
G2-M3-TE-B2-1.3.1-01.2016

Student Debrief (5 minutes)

Lesson Objective: Bundle and count ones, tens, and hundreds to 1,000.

The Student Debrief is intended to invite reflection and active processing of the total lesson experience.

Invite students to review their solutions for the Problem Set. They should check work by comparing answers with a partner before going over answers as a class. Look for misconceptions or misunderstandings that can be addressed in the Student Debrief. Guide students in a conversation to debrief the Problem Set and process the lesson.

- T: Bring your Problem Set to the carpet.
- T: Let's read our first number by units.
- S: 4 hundreds, 3 tens, 5 ones.
- T: How do we say 3 tens 5 ones?
- S: Thirty-five.
- T: We read this number as four hundred thirty-five. Say it for me.
- S: Four hundred thirty-five.
- T: How do we say the next number down?
- S: Six hundred seventy-three.
- T: Excellent. Read the next numbers on your paper to your partner. (Allow time to do so.)
- T: To begin our Problem Set, we drew two numbers. 435 is one number. 673 is another number.
- T: What are the different units in the number 435, from largest to smallest?
- S: Hundreds, tens, ones.
- T: So we used three different units to make one number!
- T: What is this unit called? (Hold up 1 hundred straws or sticks.)
- S: 1 hundred.
- T: Discuss with your partner three questions I will write on the board:
    1. How many units of 1 are in 1 ten?
    2. How many units of 10 are in 1 hundred?
    3. How many units of 100 are in **1 thousand**?
- T: I hear a lot of intelligent answers. Show me what you know by completing your Exit Ticket. Return to your seat as soon as you have it. If you finish early, count by 10 on the back of your paper as high as you can go!

Exit Ticket (3 minutes)

After the Student Debrief, instruct students to complete the Exit Ticket. A review of their work will help with assessing students' understanding of the concepts that were presented in today's lesson and planning more effectively for future lessons. The questions may be read aloud to the students.

©2015 Great Minds. eureka-math.org
G2-M3-TE-B2-1.3.1-01.2016

Name _____   Date _____

Draw models of ones, tens, and hundreds. Your teacher will tell you which numbers to model.

Lesson 1:   Bundle and count ones, tens, and hundreds to 1,000.

21

©2015 Great Minds. eureka-math.org
G2-M3-TE-B2-1.3.1-01.2016

Name _____   Date _____

1. Draw lines to match and make each statement true.

    10 tens =   1 thousand

    10 hundreds =   1 ten

    10 ones =   1 hundred

2. Circle the largest unit. Box the smallest.

    4 tens   2 hundreds   9 ones

3. Draw models of each, and label the following number.

    2 tens   7 ones   6 hundreds

EUREKA
MATH

Name _____   Date _____

1. 2 ones + _____ ones = 10   2. 6 tens + _____ tens = 1 hundred

    2 + _____ = 10   60 + _____ = 100

3. Rewrite in order from largest to smallest units.

    6 tens   Largest   _____

    3 hundreds   _____

    8 ones   Smallest   _____

4. Count each group. What is the total number of sticks in each group?

    Bundles of 100   *Bundles of 10*   *Ones*

    _____   _____   _____

    What is the total number of sticks? _____

EUREKA
MATH™

Lesson 1:   Bundle and count ones, tens, and hundreds to 1,000.

©2015 Great Minds. eureka-math.org
G2-M3-TE-B2-1.3.1-01.2016

23

5. Draw and solve.

Moses has 100 stickers. Jared has 60 stickers. Jared wants to have the same number of stickers as Moses. How many more stickers does Jared need?

Jared needs _____ more stickers.

Lesson 1:   Bundle and count ones, tens, and hundreds to 1,000.

©2015 Great Minds. eureka-math.org
G2-M3-TE-B2-1.3.1-01.2016

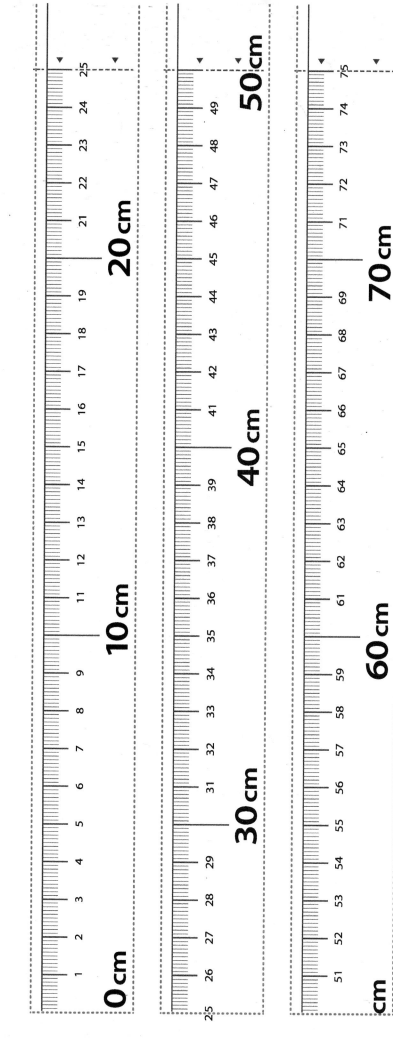

0 cm

10 cm

20 cm

30 cm

40 cm

50 cm

60 cm

70 cm

80 cm

90 cm

100 cm

cm

meter strip

LEGEND - - - - - CUT   - - - - ALIGN EDGE

Lesson 1:   Bundle and count ones, tens, and hundreds to 1,000.

©2015 Great Minds. eureka-math.org
G2-M3-TE-B2-1.3.1-01.2016

**EUREKA
MATH**

2
GRADE

Mathematics Curriculum

Topic B

Understanding Place Value Units of One, Ten, and a Hundred

2.NBT.2, 2.NBT.1

Focus Standard:	2.NBT.2	Count within 1000; skip-count by 5s, 10s, and 100s.
Instructional Days:	2	
Coherence -Links from:	G1–M6	Place Value, Comparison, Addition and Subtraction to 100
-Links to:	G2–M4	Addition and Subtraction Within 200 with Word Problems to 100

In Topic B, students practice counting by ones and skip-counting by tens and hundreds. They start off with simple counting by ones and tens in Lesson 1 (e.g., from 100 to 124 and 124 to 220). In Lesson 2, they count by ones, tens, and hundreds (e.g., from 200 to 432 and from 432 to 1,000) (**2.NBT.2**). They apply their new counting strategies to solve a *change unknown* word problem (**2.OA.1**); "Kinnear decided that he would bike 100 miles this year. If he has biked 64 miles so far, how much farther does he have to bike?"

In counting, students make use of the structure provided by multiples of 10 and 100. Students think in terms of getting to a ten or getting to a hundred. They also identify whether ones, tens, or hundreds are the appropriate unit to count efficiently and effectively. Making this determination requires knowing and understanding structures, similar to knowing the ground on which you are going to build a house and the materials with which you will build.

A Teaching Sequence Toward Mastery of Understanding Place Value Units of One, Ten, and a Hundred
Objective 1: Count up and down between 100 and 220 using ones and tens. (Lesson 2)
Objective 2: Count up and down between 90 and 1,000 using ones, tens, and hundreds. (Lesson 3)

Lesson 2

Objective: Count up and down between 100 and 220 using ones and tens.

Suggested Lesson Structure

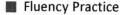

■ Fluency Practice	(18 minutes)
■ Application Problem	(8 minutes)
■ Concept Development	(24 minutes)
■ Student Debrief	(10 minutes)
Total Time	**(60 minutes)**

Fluency Practice (18 minutes)

- Meter Strip Subtraction: Taking Multiples of 10 from Numbers 10–100 **2.MD.6, 2.NBT.5**   (4 minutes)
- Measure and Compare **2.MD.4**   (6 minutes)
- Skip-Count Up and Down by Fives on the Clock **2.NBT.2**   (4 minutes)
- Counting with Ones, Tens, and Hundreds: 0 to 1,000 **2.NBT.8**   (4 minutes)

Meter Strip Subtraction: Taking Multiples of 10 from Numbers 10–100 (4 minutes)

Materials: (S) Meter strip (Lesson 1 Fluency Template)

Keep students challenged and engaged by adding a new layer of complexity to the activity in this second round. The following are suggestions for how to adapt the sequence demonstrated in Lesson 1 to match students' ability level. Suggestions are given in order from least to most complex.

- Subtract 9 and then 8 from multiples of 10 up to 100.
- Subtract any two-digit number from a multiple of 10 up to 100 (e.g., 30 – 13, 40 – 24, 60 – 45).
- Tell or write a number sentence describing sliding down from the whole amount (e.g., 50 – 10 = 40 cm).
- Create a sequence of *change unknown* slides. For example:
    - T: Start with your finger on 0. Slide up to 52 cm.
    - T: Now, slide down to 49. How many centimeters did you slide down?
    - S: 3 cm!
- Tell or write a problem to describe the *change unknown* slide (e.g., 52 cm – ____ = 49 cm).
- State that change in a sentence, including the unit (e.g., I slid down ____ centimeters).

Measure and Compare (6 minutes)

Materials: (S) Meter strip (Lesson 1 Fluency Template), personal white board

T: (Write or post the sentence frame described in the box shown to the right.) I'll name two objects, and you measure their lengths. Your goal is to determine how much longer one object is than another. Write the lengths on your personal white board so that you don't forget, and be sure to state the unit when you compare lengths.

T: Partner A, compare the lengths using the sentence frame (point to the frame).

T: Partner B, confirm that you agree with Partner A's statement. You might say, "I agree" or "I disagree." If you disagree, be sure to explain why. Each time we measure new things, switch roles.

T: Compare the length of your science book with the length of your crayon.

S: (For one minute, measure, write lengths, and compare them in partnerships.)

T: Compare the length of your desk and the length of the seat on your chair.

S: (For one minute, measure, write lengths, and compare them in partnerships.)

T: (Continue, being mindful to select objects that lead to agreement about which is longer or shorter. One student's pencil may very well be shorter than the crayon, while the other student's might be much longer.)

NOTES ON
MULTIPLE MEANS
OF ENGAGEMENT:

Encourage students to speak in complete sentences and to use academic vocabulary by writing or posting a sentence frame for this activity. The frame below exemplifies a single sentence that can be used in two scenarios.

If frames are new to students, quickly model their use, pointing to each part of the frame while speaking. Circulate as students use the frame with a partner.

The length of ___ is (more than/less than) the length of ____.

Skip-Count Up and Down by Fives on the Clock (4 minutes)

Materials: (T) "Clock" made from a 24-inch ribbon marked off at every 2 inches

In this second round, add a new layer of complexity to the work to keep students challenged and engaged. The following is a suggestion for how to adapt the vignette demonstrated in Lesson 1.

T: Skip-count by 5 until my finger stops. (Slide a finger to 4.)

S: 5, 10, 15, 20.

T: (From 4, slide a finger forward to 9.) Keep counting as I move my finger.

S: 25, 30, 35, 40, 45.

T: How many minutes have passed in all?

S: 45 minutes!

T: (Keep a finger at 9.) How many is 10 minutes less?

S: 35 minutes!

T: Good. (Put a finger back at 9.) How many is 10 minutes more?

S: 55 minutes!

©2015 Great Minds. eureka-math.org
G2-M3-TE-B2-1.3.1-01.2016

Counting with Ones, Tens, and Hundreds: 0 to 1,000 (4 minutes)

Materials: (T) Bundle of 1 hundred, 1 ten, and a single straw from Lesson 1

T: Let's count by ones, tens, and hundreds. I'll hold bundles to show you what to count by. A bundle of 100 means count by hundreds, a bundle of 10 means count by tens, and a single straw means count by ones. (Create visual support by writing the numbers on the board as students count.)

T: Let's start at 0. Ready? (Hold up a bundle of 10 until students count to 130.)

S: 10, 20, 30, 40, 50, 60, 70, 80, 90, 100, 110, 120, 130.

T: (Hold up a bundle of 100 until students count to 630.)

S: 230, 330, 430, 530, 630.

T: (Hold up a bundle of 10 until students count to 690.)

S: 640, 650, 660, 670, 680, 690.

T: (Hold up a single one until students count to 702.)

S: 691, 692, 693, 694, 695, 696, 697, 698, 699, 700, 701, 702.

T: (Isolate the numbers 698–702 by drawing a box around them.) Partner A, count these numbers up and down as fast as you can to Partner B, and then switch. If you both finish before one minute is up, try it again and see if you get faster!

Application Problem (8 minutes)

Ben and his dad have sold 60 chocolate chip cookies at the school bake sale. If they baked 100 cookies, how many cookies do they still need to sell?

T: Read this problem with me.

T: Close your eyes, and picture what you see when you hear the story.

T: Now, talk with your partner about what you can draw to solve this problem.

S: I can draw circles and put 10 in each. → It's like what we just did with the straws yesterday. → I can draw tens and count on.

T: You have two minutes to draw your picture.

S: (Draw.)

T: Explain to your partner how your drawing helps you answer the question.

T: Who would like to share his or her thinking?

NOTES ON PROBLEM SOLVING WITH RDW:

RDW stands for Read, Draw, Write— the problem solving process used throughout *A Story of Units*. Students first read for meaning. In this exemplar vignette, the teacher encourages visualization after the reading by having students close their eyes. Students should internalize the following set of questions:

- What do I see?
- Can I draw something?
- What can I draw?
- What can I learn from my drawing?

After drawing, students write a statement responding to the question.

S: I drew tens up to 100, and then I crossed off 6 tens and there were 4 left. 4 tens equals 40.
→ I drew 6 tens to show 60, and then I counted on to 100 and that was 4 more tens, so 40.
→ I drew a number bond and broke 100 into 60 and 40. → I wrote 6 + 4 = 10, so 60 + 40 = 100.
→ I drew a tape diagram. 100 is the whole and 60 is the part. Then, I wrote 60 + 40 = 100,
so 100 – 60 = 40.

T: Those are all very intelligent strategies for solving this problem! If anyone would like to add one of
these strategies to his or her paper, please do so now.

T: So, how many more cookies do Ben and his dad need to sell?

S: They need to sell 40 more cookies.

T: Let's write that statement on our paper.

Concept Development (24 minutes)

Materials: (T) 9 bundles of hundreds, 10 bundles of tens,
10 ones

Part 1: Counting from 100 to 110, 100 to 200, and 100 to 1,000.

Materials: (T) 10 ones, 10 tens, 10 hundreds

T: How many straws are in this bundle?

S: 100.

T: (Place 1 straw to students' right of the hundred.)
Now, there are one hundred one straws.

T: (Place 1 more straw to the right.) Now?

S: 102.

T: Count for me as I place units of one. (Start the count
again at 101. Then stop counting aloud as students
continue.)

S: 101, 102, 103, 104, 105, 106, 107, 108, 109, 110.

T: What unit can I make with these 10 ones?

S: 1 ten.

T: (Quickly bundle the 10 ones to make 1 ten.) Skip-count
for me as I place the units of ten. (Place tens, one at a
time, as students count.)

**NOTES ON
MULTIPLE MEANS
OF ACTION AND
EXPRESSION:**

By keeping the start number of the
count, 100, consistent, students have
the opportunity to see the difference
the units make in language patterns
and quantity.

As the teacher omits his voice in the
count, making every effort not to
mouth the numbers, students learn to
listen to their peers and to
acknowledge that their peers are
competent. Students watch the straws
but listen to the count. In doing so, the
language is associated with a quantity
as well as a sequence of number
words. This promotes retention.

If students' count is weak, have a
smaller sub-group count. "Those who
feel they know the count, try this
time." Then, have the entire group try
again. Quickly celebrate authentic
improvement.

S: 110, 120, 130, 140, 150, 160, 170, 180, 190, 200.

T: What unit can I make with these 10 tens?

S: 1 hundred.

T: (Quickly bundle the 10 tens to make 1 hundred.) Skip-count for me as I place units of 100. (Place hundreds one at a time.)

S: 100, 200, 300, 400, 500, 600, 700, 800, 900, 1,000.

T: What unit can I make with these 10 hundreds?

S: 1 thousand.

T: (Quickly bundle the 10 hundreds to make 1 thousand.)

Part 2: Counting from 100 to 124 and 124 to 100.

Materials: (T) 1 hundred, 2 tens, 4 ones (S) 1 hundred, 2 tens, 4 ones per pair

T: (Place 1 unit of 1 hundred on the carpet, but do not give students straws.) With your partner, count from 100 up to 124 using both units of one and ten.

T: (Circulate and listen. Anticipate most students will count by ones.)

T: Try again using our units. (Give each pair 1 hundred, 2 tens, and 4 ones.) Model your counting. Which is the fastest way to reach 124?

T: (Circulate and listen for, or guide, students to notice how much faster it is to count by tens than by ones up to 124.)

T: Jeremy, would you stand and show us how you use both tens and ones?

S: 100, 110, 120, 121, 122, 123, 124.

T: Alejandra, would you stand and tell us how you used both tens and ones?

S: 100, 101, 102, 103, 104, 114, 124.

T: Marco?

S: 100, 110, 111, 112, 113, 114, 124.

T: There are other ways, too. Class, please count for me Jeremy's way. (Model with the bundles as students count.)

S: 100, 110, 120, 121, 122, 123, 124.

T: Show 124 with your straws. Count down from 124 to 100. Model by taking away one unit at a time.

Part 3: Counting from 124 to 220 and 220 to 124.

Materials: (S) 9 tens and 6 ones per pair

T: (Give each pair 9 tens and 6 ones.) With your partner, count from 124 up to 220. Model with your straws as you count.

T: (Circulate and listen.)

Repeat the process from the previous count. Have students count up and down both with straws and without.

 EUREKA MATH™

Lesson 2:   Count up and down between 100 and 220 using ones and tens.

31

©2015 Great Minds. eureka-math.org
G2-M3-TE-B2-1.3.1-01.2016

Problem Set (10 minutes)

Students should do their personal best to complete the Problem Set within the allotted 10 minutes. For some classes, it may be appropriate to modify the assignment by specifying which problems they work on first. Some problems do not specify a method for solving. Students should solve these problems using the RDW approach used for Application Problems.

> T: Draw, label, and box the following numbers. (Demonstrate to the least extent possible.)
> > a.   100
> > b.   124
> > c.   85
> > d.   120
>
> T: Use both tens and ones to count up to the target numbers. Draw the tens and ones you used. Write the counting numbers.
> > a.   100 to 124
> > b.   124 to 220
> > c.   85 to 120
> > d.   120 to 193

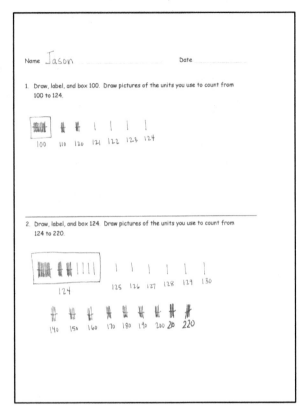

Student Debrief (10 minutes)

Lesson Objective: Count up and down between 100 and 220 using ones and tens.

Materials: (S) Straws and bundles of tens and hundreds

The Student Debrief is intended to invite reflection and active processing of the total lesson experience.

Invite students to review their solutions for the Problem Set. They should check work by comparing answers with a partner before going over answers as a class. Look for misconceptions or misunderstandings that can be addressed in the Debrief. Guide students in a conversation to debrief the Problem Set and process the lesson.

> T: I see that when Freddy counted from 124 to 220,
> MP.7 he first used ones to get to 130. Freddy, could you explain your thinking?

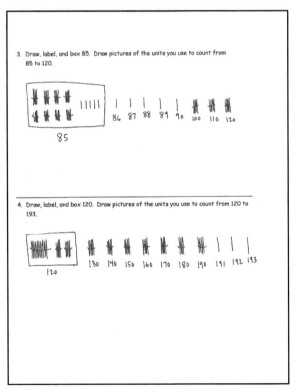

S: It's easy for me to skip-count by tens from 130, so I wanted to get to 130.

T: Freddy got to a benchmark number, 130, and then counted on. Good strategy. Share with your partner why you think I called it a benchmark number.

S: It's a benchmark number because it's helpful.
     → It makes it easier to count.

MP.7

T: You're exactly right! Benchmark numbers allow us to skip-count, which is faster than counting by ones. A bench is somewhere you sit comfortably, and so a benchmark number is something that is easy to remember and rest on.

T: Let's practice looking for benchmark numbers. Talk with your partner. What benchmark number would help you count from 85 to 120?

NOTES ON MATHEMATICAL PRACTICE 7:

Giving students opportunities to practice counting using ones and bundles of tens and hundreds while asking them to identify benchmark numbers will cue them to the ease and efficiency of skip-counting.

It will accustom them to look for, and make use of, the structure provided by the base ten number system, not only to skip-count from multiples of ten but also multiples of 100, and later, larger units.

As students talk, circulate, listen, and support. Decide on whom to call to report out to the class.)

T: Monica, could you please use the straws and bundles to demonstrate?

S: I used ones to count up to 90 and then counted by tens to 120.

T: What was the benchmark number Monica got to?

S: 90.

T: Let's count as Monica shows us again.

S: 85, 86, 87, 88, 89, 90.

T: Stop. Why is 90 a benchmark number? How does 90 help us?

S: Now we can skip-count by 10, which is faster.

T: Yes!

T: Let's try another one. What benchmark number would you use if you were counting from 156 to 200?

S: 160.

T: George, could you please show us with the straws as we count?

S: 156, 157, 158, 159, 160.

T: Now, what unit will we count by?

S: Tens!

T: Let's hear it!

S: 170, 180, 190, 200.

T: What benchmark number would you use if you were counting from 97 to 200?

S: 100.

T: Sometimes even a benchmark number needs help. If I'm counting from 70 to 200, what benchmark number do I want to get to? Talk to your partner.

S: 100.

T: What unit did you use to get to 100?

S: Tens.

T: What unit did you use to count from 100 to 200?

S: Hundreds!

MP.7

T: What about if I'm counting from 76 to 200? What units would I use? Talk with your partner.

S: Ones, tens, and hundreds!

T: I'll place the straws and bundles as you count. Go!

S: 76, 77, 78, 79, 80, 90, 100, 200.

T: Benchmark numbers are structures that help us count up and down. We can use both different units and benchmark numbers to make counting easier.

Exit Ticket (3 minutes)

After the Student Debrief, instruct students to complete the Exit Ticket. A review of their work will help with assessing students' understanding of the concepts that were presented in today's lesson and planning more effectively for future lessons. The questions may be read aloud to the students.

©2015 Great Minds. eureka-math.org
G2-M3-TE-B2-1.3.1-01.2016

Name _____ Date _____

1. Draw, label, and box 100. Draw pictures of the units you use to count from 100 to 124.

_____

2. Draw, label, and box 124. Draw pictures of the units you use to count from 124 to 220.

Lesson 2:   Count up and down between 100 and 220 using ones and tens.

35

©2015 Great Minds. eureka-math.org
G2-M3-TE-B2-1.3.1-01.2016

3. Draw, label, and box 85. Draw pictures of the units you use to count from 85 to 120.

4. Draw, label, and box 120. Draw pictures of the units you use to count from 120 to 193.

EUREKA
MATH

©2015 Great Minds. eureka-math.org
G2-M3-TE-B2-1.3.1-01.2016

Name _____   Date _____

1. These are bundles of hundreds, tens, and ones. How many straws are in each group?

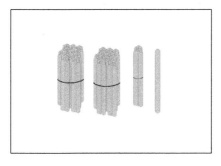   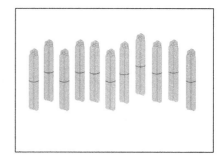

_____ straws   _____ straws

2. Count from 96 to 140 with ones and tens. Use pictures to show your work.

3. Fill in the blanks to reach the benchmark numbers.

35, _____, _____, _____, _____, 40, _____, _____, _____, _____, _____, 100, _____, 300

©2015 Great Minds. eureka-math.org
G2-M3-TE-B2-1.3.1-01.2016

Name _____   Date _____

1. How many in all?

   _____ ones = _____ tens

   _____ stars in all.

2. These are bundles with 10 sticks in each.

   a. How many tens are there? _____

   b. How many hundreds? _____

   c. How many sticks in all? _____

3. Sally did some counting. Look at her work. Explain why you think Sally counted this way.

   177, 178, 179, 180, 190, 200, 210, 211, 212, 213, 214

EUREKA
MATH™

4. Show a way to count from 68 to 130 using tens and ones. Explain why you chose to count this way.

5. Draw and solve.

    In her classroom, Sally made 17 bundles of 10 straws. How many straws did she bundle in all?

©2015 Great Minds. eureka-math.org
G2-M3-TE-B2-1.3.1-01.2016

Lesson 3

Objective: Count up and down between 90 and 1,000 using ones, tens, and hundreds.

Suggested Lesson Structure

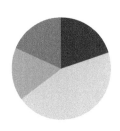

■ Fluency Practice   (12 minutes)
　Concept Development   (27 minutes)
■ Application Problem   (10 minutes)
■ Student Debrief   (11 minutes)

Total Time   **(60 minutes)**

Fluency Practice (12 minutes)

- Sprint: Differences to 10 with Teen Numbers **2.OA.2**   (8 minutes)
- Counting with Ones, Tens, and Hundreds from 0 to 1,000 **2.NBT.8**   (4 minutes)

Sprint: Differences to 10 with Teen Numbers (8 minutes)

Materials: (S) Differences to 10 with Teen Numbers Sprint

Counting with Ones, Tens, and Hundreds from 0 to 1,000 (4 minutes)

Materials: (T) Bundle of 1 hundred, 1 ten, and a single straw from Lesson 1

For this second round, you may want to change the partner share to have students rapidly count up and down a larger sequence of numbers. Students often need additional practice with crossing a hundred, as well as with the first 30 numbers that begin a new hundred (e.g., 100–130, 600–630).

NOTES ON LESSON STRUCTURE:

Application Problems can follow Concept Development so that students connect and apply their learning to real-world situations. They can also serve as a lead-in to a concept, allowing students to discover through problem solving the logic and usefulness of a strategy before that strategy is formally taught. This gives students a framework on which to hang their developing understanding.

©2015 Great Minds. eureka-math.org
G2-M3-TE-B2-1.3.1-01.2016

Concept Development (27 minutes)

Materials: (T) 9 units of 1 hundred, 10 units of ten, 10 ones
(for Parts A, B, C, and D)

Part A

Part A Sequence
Count from
90 to 300
170 to 500
350 to 600
780 to 1,000

NOTES ON
MULTIPLE MEANS
OF ACTION AND
EXPRESSION:

Students working above grade level
may combine Parts A and B, then C and
D. Challenge students to count from
90 to 300 to 480.

For struggling students, adjust the task
such that they only complete Parts A
and B. The rest can be practiced
during fluency time throughout the
year. To ease students into counting
without physical units, model with the
straws and then hide them under a
sheet of paper. Prompt students to
visualize as they count.

T: Today, let's use units of ten and a hundred to count
from 90 to 300. (Place 9 units of ten on the carpet.)

T: I'll model. You count. (Place bundles as they count.)

S: 90, 100, 200, 300.

T: Now, let's count down from 300 to 90.

S: 300, 200, 100, 90. (Remove bundles as they count.)

T: Talk to your partner about how we counted up and down.

S: First, put 1 ten to get to a benchmark number, 100. Then, keep counting by hundreds. 200, 300.

Quickly do further examples from the Part A Sequence chart. Students get very excited about the larger
numbers.

T: Is it faster to count using tens or hundreds?

S: Hundreds.

T: Why?

S: They are bigger, so you get there faster. → It's like you don't have to say as many numbers.
→ If you don't know how to count by hundreds, it might be faster to count by tens.

If necessary, have students practice using their own bundles with small amounts such as 90 to 200, 80 to 200,
and 60 to 300.

Part B

Part B Sequence
Count from
300 to 480
500 to 830
600 to 710
800 to 990

Next, count between pairs of numbers starting with multiples of 100 and ending with numbers that have both
hundreds and tens, such as 300 to 480, as shown in the Part B Sequence chart.

Lesson 3:   Count up and down between 90 and 1,000 using ones, tens, and
hundreds.

41

Parts C and D

Part C Sequence	Part D Sequence
Count from	Count from
100 to 361	361 to 400
200 to 432	432 to 600
600 to 725	725 to 900
700 to 874	874 to 1,000

Advance to using 3 units while counting up and down between pairs of numbers.

Problem Set (10 minutes)

Students should do their personal best to complete the Problem Set within the allotted 10 minutes. For some classes, it may be appropriate to modify the assignment by specifying which problems they work on first. Some problems do not specify a method for solving. Students should solve these problems using the RDW approach used for Application Problems.

T: Draw, label, and box the following numbers. (Demonstrate to the least extent possible).

    a.   90

    b.   300

    c.   428

    d.   600

T: Draw pictures of the units you use to count up to the target number. Use hundreds whenever you can, or you won't have space on your paper.

    a.   90 to 300

    b.   300 to 428

    c.   428 to 600

    d.   600 to 1,000

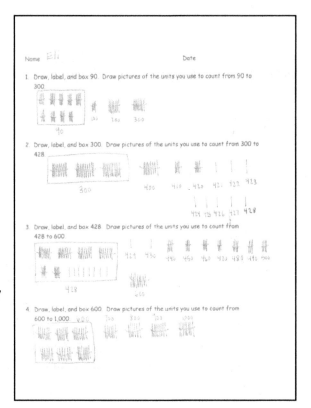

NOTES ON MULTIPLE MEANS OF REPRESENTATION:

Even the simplest illustration brings a story to life, especially for English language learners. Draw a bicycle and a road. Add a sign post. Replace an unfamiliar name like Kinnear with a name from the class. Allow students to use a set of bundles if they choose. Then, have them return to their seats and draw.

As often as possible, invite students to show their work while talking about it. Have them point to the places they are referring to in their counting sequence. This visual input is perfect for English language learners and students performing below grade level because it keeps them focused on sense-making.

Lesson 3:   Count up and down between 90 and 1,000 using ones, tens, and hundreds.

©2015 Great Minds. eureka-math.org
G2-M3-TE-B2-1.3.1-01.2016

Application Problem (10 minutes)

Kinnear decided that he would bike 100 miles this year. If he has biked 64 miles so far, how much farther does he have to bike?

T: Let's read the problem.

T: Talk with your partner: Do we know the parts, or do we know the whole and one part?

S: We know the whole and one part.

T: Which means we're looking for…?

S: The missing part!

T: Tell your partner the subtraction problem that goes with this story. Raise your hand when you know the answer.

S: 100 − 64 = blank.

T: Talk with your partner: What is a related addition fact?

S: 64 + blank = 100.

T: Draw a picture to show how you can use units of one and ten to find the answer. You have two minutes.

S: 70 was my benchmark number. I drew 6 ones to get to 70. Then I drew 3 tens to make 1 hundred.

T: Let's count using Jorge's model.

S: 65, 66, 67, 68, 69, 70, 80, 90, 100.

T: Did anyone use a different counting strategy?

S: I counted by tens from 64 to 94 and that was 3 tens; then I added 6 ones to make 100.

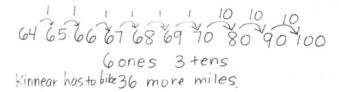

T: So, if we count Jorge's way, we add 6 ones and 3 tens, which equals…?

S: 36.

T: And, if we add Delilah's way, we add 3 tens and 6 ones, which equals…?

S: 36.

T: Are both counting strategies correct?

S: Yes!

T: So, how much farther does Kinnear have to bike?

S: Kinnear has to bike 36 more miles.

T: Add that sentence to your paper.

Lesson 3:   Count up and down between 90 and 1,000 using ones, tens, and
                  hundreds.

©2015 Great Minds. eureka-math.org
G2-M3-TE-B2-1.3.1-01.2016

43

Student Debrief (11 minutes)

Lesson Objective: Count up and down between 90 and 1,000 using ones, tens, and hundreds.

The Student Debrief is intended to invite reflection and active processing of the total lesson experience.

Invite students to review their solutions for the Problem Set. They should check work by comparing answers with a partner before going over answers as a class. Look for misconceptions or misunderstandings that can be addressed in the Debrief. Guide students in a conversation to debrief the Problem Set and process the lesson.

Students bring their Problem Set and Application Problem solutions.

T: Let's look at the first count you did from 90 to 300.

T: What was your first benchmark number?

S: 100.

T: How many tens did we count to get there?

S: 1 ten.

T: How many hundreds did we count to get from 100 to 300?

S: 2 hundreds.

T: So, in all how much did we count to get from 90 to 300?

S: 1 ten and 2 hundreds.

T: How much is that in all?

S: 210.

T: Where is 210 on your Problem Set?

S: It's the part that isn't boxed right here.

T: So, how many straws are in this part?

S: 90 straws.

T: How many straws are in this part?

S: 210 straws.

T: When you put them together, let's count what we get. (Touch as students count.)

S: 90, 100, 200, 300. 300 straws!

T: Talk to your partner: Can your counting help you to solve the problem about Kinnear?

MP.7

S: I thought that every straw was 1 mile. → It was like counting up. → I started at 64 and added ones to get to 70. → 65, 66, 67, 68, 69, 70. That is, 1, 2, 3, 4, 5, 6 ones. → 80, 90, 100. That is 1, 2, 3 tens. → 6 ones and 3 tens is 36.

T: So, what does 36 mean to Kinnear?

S: That's how many miles he has to go.

T: Look at Problem 4. Suppose Kinnear has gone 600 miles. How many miles does he have to go to bike 1,000 miles?

S: 400 miles!

Lesson 3:   Count up and down between 90 and 1,000 using ones, tens, and hundreds.

©2015 Great Minds. eureka-math.org
G2-M3-TE-B2-1.3.1-01.2016

T: What if Kinnear had only gone 90 miles? How far would he still have to go to bike 1,000 miles? Talk to your partner. (Model only the units necessary for the count.)

S: 100, 200, 300, …, 900, 1,000.

T: Work with your partner. How many straws do you see we counted? (Be sure they are easy to see.)

S: 910.

T: What units did you use?

S: A ten and 9 hundreds.

T: That is the part we needed to get from 90 to 1,000.

T: Tell me which unit or units to use: ones, tens, or hundreds? (Pause after each of the following questions.)

MP.7

T: To count from 36 to 40?

S: Ones!

T: To count from 36 to 100?

S: Ones and tens!

T: To count from 100 to 800?

S: Hundreds!

T: To count from 70 to 100?

S: Tens.

T: To get from 67 to 600?

S: Ones, tens, and hundreds!

Exit Ticket (3 minutes)

After the Student Debrief, instruct students to complete the Exit Ticket. A review of their work will help with assessing students' understanding of the concepts that were presented in today's lesson and planning more effectively for future lessons. The questions may be read aloud to the students.

EUREKA MATH™

©2015 Great Minds. eureka-math.org
G2-M3-TE-B2-1.3.1-01.2016

A

Number Correct: _____

Differences to 10 with Teen Numbers

1.	3 – 1 =		23.	7 – 4 =	
2.	13 – 1 =		24.	17 – 4 =	
3.	5 – 1 =		25.	7 – 5 =	
4.	15 – 1 =		26.	17 – 5 =	
5.	7 – 1 =		27.	9 – 5 =	
6.	17 – 1 =		28.	19 – 5 =	
7.	4 – 2 =		29.	7 – 6 =	
8.	14 – 2 =		30.	17 – 6 =	
9.	6 – 2 =		31.	9 – 6 =	
10.	16 – 2 =		32.	19 – 6 =	
11.	8 – 2 =		33.	8 – 7 =	
12.	18 – 2 =		34.	18 – 7 =	
13.	4 – 3 =		35.	9 – 8 =	
14.	14 – 3 =		36.	19 – 8 =	
15.	6 – 3 =		37.	7 – 3 =	
16.	16 – 3 =		38.	17 – 3 =	
17.	8 – 3 =		39.	5 – 4 =	
18.	18 – 3 =		40.	15 – 4 =	
19.	6 – 4 =		41.	8 – 5 =	
20.	16 – 4 =		42.	18 – 5 =	
21.	8 – 4 =		43.	8 – 6 =	
22.	18 – 4 =		44.	18 – 6 =	

Lesson 3: Count up and down between 90 and 1,000 using ones, tens, and hundreds.

EUREKA MATH

©2015 Great Minds. eureka-math.org
G2-M3-TE-B2-1.3.1-01.2016

B

Number Correct: _____

Improvement: _____

Differences to 10 with Teen Numbers

1.	2 – 1 =	
2.	12 – 1 =	
3.	4 – 1 =	
4.	14 – 1 =	
5.	6 – 1 =	
6.	16 – 1 =	
7.	3 – 2 =	
8.	13 – 2 =	
9.	5 – 2 =	
10.	15 – 2 =	
11.	7 – 2 =	
12.	17 – 2 =	
13.	5 – 3 =	
14.	15 – 3 =	
15.	7 – 3 =	
16.	17 – 3 =	
17.	9 – 3 =	
18.	19 – 3 =	
19.	5 – 4 =	
20.	15 – 4 =	
21.	7 – 4 =	
22.	17 – 4 =	

23.	9 – 4 =	
24.	19 – 4 =	
25.	6 – 5 =	
26.	16 – 5 =	
27.	8 – 5 =	
28.	18 – 5 =	
29.	8 – 6 =	
30.	18 – 6 =	
31.	9 – 6 =	
32.	19 – 6 =	
33.	9 – 7 =	
34.	19 – 7 =	
35.	9 – 8 =	
36.	19 – 8 =	
37.	8 – 3 =	
38.	18 – 3 =	
39.	6 – 4 =	
40.	16 – 4 =	
41.	9 – 5 =	
42.	19 – 5 =	
43.	7 – 6 =	
44.	17 – 6 =	

EUREKA MATH

Lesson 3: Count up and down between 90 and 1,000 using ones, tens, and hundreds.

47

©2015 Great Minds. eureka-math.org
G2-M3-TE-B2-1.3.1-01.2016

Name _____ Date _____

1. Draw, label, and box 90. Draw pictures of the units you use to count from 90 to 300.

2. Draw, label, and box 300. Draw pictures of the units you use to count from 300 to 428.

Lesson 3:   Count up and down between 90 and 1,000 using ones, tens, and hundreds.

©2015 Great Minds. eureka-math.org
G2-M3-TE-B2-1.3.1-01.2016

EUREKA
MATH™

3. Draw, label, and box 428. Draw pictures of the units you use to count from 428 to 600.

4. Draw, label, and box 600. Draw pictures of the units you use to count from 600 to 1,000.

 Lesson 3: Count up and down between 90 and 1,000 using ones, tens, and hundreds.

©2015 Great Minds. eureka-math.org
G2-M3-TE-B2-1.3.1-01.2016

49

Name _____   Date _____

1. Draw a line to match the numbers with the units you might use to count them.

    300 to 900   ones, tens, and hundreds

    97 to 300   ones and tens

    484 to 1,000   ones and hundreds

    743 to 800   hundreds

2. These are bundles of hundreds, tens, and ones. Draw to show how you would count to 1,000.

Lesson 3:   Count up and down between 90 and 1,000 using ones, tens, and
                            hundreds.

EUREKA
MATH™

Name _____   Date _____

1. Fill in the blanks to reach the benchmark numbers.

   a. 14, _____, _____, _____, _____, _____, 20, _____, _____, 50

   b. 73, _____, _____, _____, _____, _____, _____, 80, _____, 100, _____, 300, _____, 320

   c. 65, _____, _____, _____, _____, 70, _____, _____, 100

   d. 30, _____, _____, _____, _____, _____, _____, 100, _____, _____, 400

2. These are ones, tens, and hundreds. How many sticks are there in all?

   There are _____ sticks in all.

3. Show a way to count from 668 to 900 using ones, tens, and hundreds.

EUREKA
MATH™

Lesson 3:   Count up and down between 90 and 1,000 using ones, tens, and
hundreds.

51

©2015 Great Minds. eureka-math.org
G2-M3-TE-B2-1.3.1-01.2016

4. Sally bundled her sticks in hundreds, tens, and ones.

   a. How many sticks does Sally have? _____

   b. Draw 3 more hundreds and 3 more tens. Count and write how many sticks Sally has now.

Lesson 3:   Count up and down between 90 and 1,000 using ones, tens, and hundreds.

EUREKA
MATH™

Mathematics Curriculum

Topic C
Three-Digit Numbers in Unit, Standard, Expanded, and Word Forms

2.NBT.3, 2.NBT.1, 2.NBT.2

Focus Standard:	2.NBT.3	Read and write numbers to 1000 using base-ten numerals, number names, and expanded form.
Instructional Days:	2	
Coherence -Links from:	G1–M6	Place Value, Comparison, Addition and Subtraction to 100
-Links to:	G2–M4	Addition and Subtraction Within 200 with Word Problems to 100
	G2–M7	Problem Solving with Length, Money, and Data

In Topic C, the teaching sequence opens with students counting on the place value chart by ones from 0 to 124, bundling larger units as possible (**2.NBT.1a**). Next, they represent various counts in numerals, also known as standard form, designating and analyzing benchmark numbers (e.g., multiples of 10) and numbers they bundled to count by a larger unit (**2.NBT.2**).

Next, students work with base ten numerals representing modeled numbers with place value cards, also known as Hide Zero cards, that reveal or hide the value of each place. They represent three-digit numbers as number bonds and gain fluency in expressing numbers in unit form (3 hundreds 4 tens 3 ones), in word form, and on the place value chart. Students then count up by hundreds, tens, and ones, leading them to represent numbers in expanded form (**2.NBT.3**). The commutative property or *switch around rule* allows them to change the order of the units. They practice moving fluidly between word form, unit form, standard form, and expanded form (**2.NBT.3**).

Students are held accountable for naming the unit they are talking about, manipulating, or counting. Without this precision, they run the risk of thinking of numbers as simply the compilation of numerals 0–9, keeping their number sense underdeveloped.

The final Application Problem involves a found briefcase full of money: 23 ten-dollar bills, 2 hundred-dollar bills, and 4 one-dollar bills. Students use both counting strategies and place value knowledge to find the total value of the money.

A Teaching Sequence Toward Mastery of Three-Digit Numbers in Unit, Standard, Expanded, and Word Forms

Objective 1: Count up to 1,000 on the place value chart.
(Lesson 4)

Objective 2: Write base ten three-digit numbers in unit form; show the value of each digit.
(Lesson 5)

Objective 3: Write base ten numbers in expanded form.
(Lesson 6)

Objective 4: Write, read, and relate base ten numbers in all forms.
(Lesson 7)

EUREKA
MATH™

Lesson 4

Objective: Count up to 1,000 on the place value chart.

Suggested Lesson Structure

- ■ Fluency Practice   (15 minutes)
- ▨ Application Problem   (7 minutes)
- ▨ Concept Development   (28 minutes)
- ■ Student Debrief   (10 minutes)

   Total Time   **(60 minutes)**

Fluency Practice (15 minutes)

- Sprint: Adding to the Teens **2.OA.2**   (10 minutes)
- Exchange to Get to 50 **2.NBT.2**   (5 minutes)

Sprint: Adding to the Teens (10 minutes)

Materials: (S) Adding to the Teens Sprint

Exchange to Get to 50 (5 minutes)

Materials: (S) Dienes blocks: 12 ones, 5 tens, and 1 hundred; 1 die per pair

Suggestions for modifying this game are presented in Lesson 5.

   T: Working with your partner, our goal is to make 50.
   T: Partner A, roll the die. Take that number of ones cubes from your pile, and line them up in the first row on your hundreds flat.
   T: Now Partner B takes a turn.
   T: It's Partner A's turn again. Start a new row if you need to.
   T: Some of you may now have 10, 11, or 12 ones on your hundred flats. If you completed a ten with your last roll, exchange the row of 10 ones for a tens rod. Be sure to leave your extra ones on your hundreds flat.
   T: Now it's Partner B's turn. Keep taking turns until the first person reaches 50.

NOTES ON DIENES BLOCKS:

These are often called base ten blocks. Dienes blocks include hundreds flats, tens rods, and ones cubes. They have not been formally introduced in Grade 2, and many students find them difficult to use at first. It may be appropriate to briefly identify each type of block before starting the game.

However, avoid taking time to teach to the manipulative. The game itself provides students with the opportunity to explore the blocks and their relationship to one another.

Application Problem (7 minutes)

At his birthday party, Joey got $100 from each of his two grandmothers, $40 from his dad, and $5 from his little sister. How much money did Joey get for his birthday?

T: Read this problem with me.

T: Take a minute to talk with your partner about what information this problem gives you and how you can draw it.

T: (Circulate and listen for sound reasoning but also for common misperceptions.)

S: I can show $100 and $40 and $5.

T: Does anyone disagree with what Susana said? If so, can you explain why?

S: Each grandma gave Joey $100, and Joey has 2 grandmas, so it's $200, not $100.

T: Yes. It's very important to read carefully. Now draw your pictures and solve.

T: (After a minute or two.) Let's use Elijah's drawing to count and find the answer.

S: 100, 200, 210, 220, 230, 240, 241, 242, 243, 244, 245.

T: 245 what?

S: 245 dollars!

T: Give me the statement.

S: Joey got $245 for his birthday.

T: Talk with your partner. What does counting money remind you of? It's like counting…?

S: Hundreds, tens, and ones!

T: How many of each unit are in $245?

S: 2 hundreds, 4 tens, 5 ones.

T: Very well done. Please write the answer, "Joey got $245 for his birthday." on your paper.

NOTES ON MAINTAINING COHERENCE WITHIN THE LESSON:

This problem could be solved in multiple ways. Resist the temptation to use or show expanded form to solve this problem. Students may come up with it; however, our intent here is to stay focused on counting as an addition strategy as modeled in the vignette.

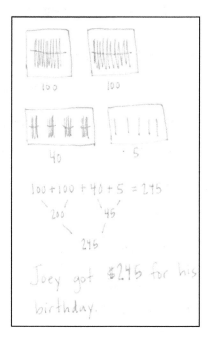

EUREKA
MATH™

Concept Development (28 minutes)

Count Up to 10 by Ones, to 100 by Tens, and to 1,000 by Hundreds on the Place Value Chart (3 minutes)

Materials: (T) 3 shoe box lids joined to create a place value "box" labeled *hundreds, tens, and ones*; Hide Zero cards (Template 1); 10 straws; bundles of tens and hundreds from Lesson 1; rubber bands (S) About 150 straws, 16 rubber bands, hundreds place value chart (Template 2) per pair; personal white board per student

T: (Show 1 straw.) This is 1 one. (Put the Hide Zero card in front of the box.)

T: Let's count more ones into my place value box. Count the ones with me.

S: 1 one, 2 ones, 3 ones, 4 ones, 5 ones, 6 ones, 7 ones, 8 ones, 9 ones.

T: Wait! If I put another one I can make a larger unit! What will that new, larger unit be?

S: 1 ten.

T: Let's make 1 ten. (Complete the ten, bundle it and place it into the second box.) Now how many ones are in my ones box?

Count 10 ones	1	2	3	4	5	6	7	8	9	10
Count 10 tens	10	20	30	40	50	60	70	80	90	100
Count 10 hundreds	100	200	300	400	500	600	700	800	900	1000

S: 0 ones.

T: How many tens are in my tens box?

S: 1 ten.

T: (Show the corresponding Hide Zero card and point.) 1 ten, 0 ones.

T: Let's count more tens into my place value box. Count the tens with me.

S: 1 ten, 2 tens, 3 tens, 4 tens, 5 tens, 6 tens, 7 tens, 8 tens, 9 tens.

T: Wait! If I put another unit of ten I can make a larger unit! What will that new larger unit be?

S: 1 hundred!

T: Let's make 1 hundred. (Complete the hundred, bundle it, and place it into the third box.) Now how many tens are there in my tens box?

S: 0 tens.

T: How many ones are in my ones box?

S: 0 ones!

T: (Show them the corresponding Hide Zero card and point.) 1 hundred, 0 tens, 0 ones.

(Repeat the process with hundreds.)

1 0

NOTES ON
MULTIPLE MEANS
OF ACTION AND
EXPRESSION:

At first, it may be wise to post a chart such as that pictured above. Read each row from left to right so that students see the number form as you say the unit form of the count.

1 0 0

Students Count by Ones from 0 to 124 While Bundling Units on the Place Value Chart (8 minutes)

T: Here is your **place value chart** and some straws. With your partner, I want you to count at least from 0 to 124 by ones. Whisper count while using your place value chart. Bundle a larger unit when you can.

T: What number will you show and say after 10?

S: 20.

T: No, that is what we did together. You are counting by units of one. What number will you show and say after 10?

S: 11.

T: Good. Change who places the straws each time you make ten. You have 5 minutes. (Circulate and encourage them to count out loud as they bundle tens and place them in the correct place. Work until each pair has at least counted to 124. Encourage them at times to count in unit form, at times with numerals.) Early finishers go beyond 124.

T: Now, count up to 124 on your place value chart using all three units: ones, tens, and hundreds.

T: (Model.) 1 hundred, 1 hundred ten, 1 hundred twenty, … .

T: That was a lot faster!

T: Who remembers the word that means fast and accurate?

S: Efficient!

T: That's right!

**NOTES ON
WHY 124?**

Even older students often count incorrectly 119, 120, 200. Also, Grade 1 standards count up to 120.
Be aware that the count to 124 will be used in the Student Debrief. Later in the lesson, students count from 476 to 600 and discover in the Student Debrief that the missing part is 124. They then compare the way they counted from 0 to 124 to the way they counted from 476 to 600.

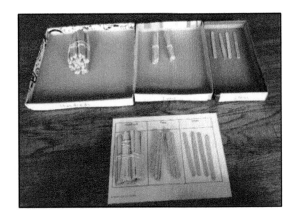

Counting Up with the Place Value Chart (7 minutes)

MP.6

T: Now, let's count today from 476 to 600 using my place value box. (Model 476 using the shoe boxes and bundles as illustrated.)

T: Let's analyze 476. How many hundreds do you see?

S: 4.

T: Tell me the unit.

S: 4 hundreds.

T: How many tens do you see?

S: 7 tens.

T: How many ones do you see?

S: 6 ones.

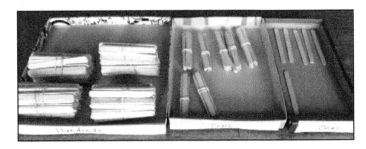

©2015 Great Minds. eureka-math.org
G2-M3-TE-B2-1.3.1-01.2016

T: We want to count from 476 to 600. Let's not count by ones. Instead let's wisely use ones, tens, and hundreds.

T: Talk to your partner about what benchmark numbers to get to and what units to use to get there.

S: Use ones to get to 480. Then use tens to get to 500. Then use a hundred to get to 600. → Count 1 hundred to get to 576. Then count ones to get to 580. Then count tens to reach 600. → Count tens to get to 496. Count ones to get to 500. Then count 1 hundred to get to 600.

MP.6

(Circulate and support students in targeting each benchmark number and each unit.)

T: Let's try it. What unit will I count first?

S: Ones.

T: Up to what benchmark number?

S: 80.

T: Really? This number is much larger than 80.

S: 480.

T: Count for me. (Place ones.)

S: 477, 478, 479, 480.

T: What do I do now?

S: Bundle a ten.

T: Now, what unit will I count by?

S: Tens!

T: Up to what benchmark number?

S: 500.

T: Count for me. (Place tens as students count.)

S: 490, 500.

T: What do I do now?

S: Bundle 1 hundred made from your 10 tens.

T: Now, what unit will I count by?

S: Hundreds.

T: Count for me. (Place 1 hundred.)

S: 600.

T: Discuss with your partner how we counted from 476 to 600 on the place value chart. Be sure to talk about the units you used, your benchmark numbers, and your bundling.

NOTES ON
MULTIPLE MEANS
OF ENGAGEMENT:

For this activity and while completing the Problem Set, work with a small group of struggling students. The smaller group setting and the use of large manipulatives (e.g., teacher place value boxes) supports students as they count up from one number to the next and as they move back and forth between unit form and numerals. Encourage them to use the stems, "I can change 10 ones for 1 ten," and "I can change 10 tens for 1 hundred."

S: (Share with partners.)

MP.6

T: Can you write the numbers that tell the way you counted? Let's start with 477.

S: 477, 478, 479, 480, 490, 500, 600.

T: Let's underline where we bundled a larger number and where we got to a benchmark number.

Problem Set (10 minutes)

Students should do their personal best to complete the Problem Set within the allotted 10 minutes. For some classes, it may be appropriate to modify the assignment by specifying which problems they work on first. Some problems do not specify a method for solving. Students should solve these problems using the RDW approach used for Application Problems.

Complete Problem 1 on the Problem Set as a guided practice with the class before allowing students to continue with Problems 2–4.

- Problem 1   476 to 600   (guided practice)
- Problem 2   47 to 200
- Problem 3   188 to 510
- Problem 4   389 to 801

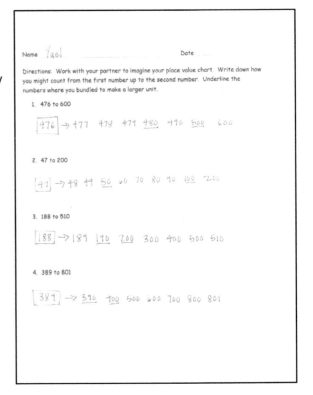

Student Debrief (10 minutes)

Lesson Objective: Count up to 1,000 on the place value chart.

Materials: (T) 3 shoe box lids joined to create a place value "box" labeled *hundreds, tens, and ones*; 10 straws; bundles of tens and hundreds from Lesson 1; rubber bands (S) Completed Problem Set

The Student Debrief is intended to invite reflection and active processing of the total lesson experience.

Invite students to review their solutions for the Problem Set. They should check work by comparing answers with a partner before going over answers as a class. Look for misconceptions or misunderstandings that can be addressed in the Debrief. Guide students in a conversation to debrief the Problem Set and process the lesson.

T: Bring your work to the carpet. Talk to your elbow partner. Where did you bundle a new unit in each count?

S: (Share.)

T: Let's prove your thoughts by modeling each count quickly on the **place value chart**. Let's start with Problem 2, 47 to 200.

T: Count while I place the straws. Tell me what to bundle when necessary.

EUREKA
MATH

S: 48, 49, 50. Bundle a ten! 60, 70, 80, 90, 100. Bundle a hundred. 200.

T: Problem 3, 188 to 510. Count while I place the straws. Tell me what to bundle when necessary.

S: 189, 190. Bundle a ten! 200. Bundle a hundred. 300, 400, 500, 510.

T: Problem 4, 389 to 801. (Model as students count.)

S: 390. Bundle a ten! 400. Bundle a hundred! 500, 600, 700, 800, 801.

T: Good. Let's take a look at something interesting.

T: (Place 476 to the side of the place value box.) Let's count from 476 to 600 again, but this time let's only show what we are counting in our place value box.

T: What unit did we start our count with?

S: Ones.

T: Count for me. (Place straws in the appropriate box as students count.)

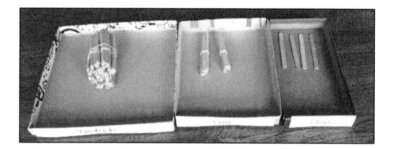

S: 477, 478, 479, 480.

T: Keep going.

S: 490, 500.

T: Keep going.

S: 600.

T: What is the value of what I counted on from 476 to get to 600?

S: 124. That is the same number we counted before!

T: Yes! We did count from 0 to 124 first. Next, we counted from 476 to 600.

T: Talk to your partner. How is counting from 0 to 124 different from this new 124 we found when we counted from 476 to 600?

S: We got to a much bigger number! → Before we started at 0. This time we started at 476. → We counted the hundred first before, but this time we counted the ones first. → Going to 124, our first benchmark number was 100. This time it was 480. → And we didn't have to bundle counting to 124, but we did counting to 600. → It's because the first time we counted to 124, the ones came last. This time they came first. → It's like this was the part that was missing to get from 476 to 600. → Yes, this is the missing part. We filled in the 24 first to get to 500 and then added the hundred.

T: I would like to hear some thoughts from the people I spoke to while you were partner sharing.

T: Jessica and Orlando, would you share?

S: We noticed that from 0 to 124 and from 476 to 600, there are 124 between both of them when you count up.

T: Who can rephrase that in their own words?

S: It takes as many straws to get from 0 to 124 as it takes to get from 476 to 600.

T: Yes. Can someone use the words *missing part* to restate the same idea?

S: It is the same missing part, 124, to count from 0 to 124 and from 476 to 600.

T: Turn and talk to your partner about what your friends noticed.

T: Do you think there are other pairs of numbers like 476 and 600 where the count is 124 between them?

©2015 Great Minds. eureka-math.org
G2-M3-TE-B2-1.3.1-01.2016

S: Yes!

T: Think about it during the week. On Friday, if anyone wants to share another pair of numbers, we would love to hear them. Talk to your family members about it, too.

Exit Ticket (3 minutes)

After the Student Debrief, instruct students to complete the Exit Ticket. A review of their work will help with assessing students' understanding of the concepts that were presented in today's lesson and planning more effectively for future lessons. The questions may be read aloud to the students.

©2015 Great Minds. eureka-math.org
G2-M3-TE-B2-1.3.1-01.2016

A

Number Correct: _____

Adding to the Teens

1.	5 + 5 + 5 =		23.	1 + 9 + 5 =	
2.	9 + 1 + 3 =		24.	3 + 5 + 5 =	
3.	2 + 8 + 4 =		25.	8 + 4 + 6 =	
4.	3 + 7 + 2 =		26.	9 + 7 + 1 =	
5.	4 + 6 + 9 =		27.	2 + 6 + 8 =	
6.	9 + 0 + 6 =		28.	0 + 8 + 7 =	
7.	3 + 0 + 8 =		29.	8 + 4 + 3 =	
8.	2 + 7 + 7 =		30.	9 + 2 + 2 =	
9.	6 + 6 + 6 =		31.	4 + 4 + 4 =	
10.	7 + 8 + 4 =		32.	6 + 8 + 5 =	
11.	3 + 5 + 9 =		33.	4 + 5 + 7 =	
12.	9 + 1 + 1 =		34.	7 + 3 + 1 =	
13.	5 + 5 + 6 =		35.	6 + 4 + 3 =	
14.	8 + 2 + 8 =		36.	1 + 9 + 9 =	
15.	3 + 4 + 7 =		37.	5 + 8 + 5 =	
16.	5 + 0 + 8 =		38.	3 + 3 + 5 =	
17.	6 + 2 + 6 =		39.	7 + 0 + 6 =	
18.	6 + 3 + 9 =		40.	4 + 5 + 9 =	
19.	2 + 4 + 7 =		41.	4 + 8 + 4 =	
20.	3 + 8 + 6 =		42.	2 + 6 + 7 =	
21.	5 + 7 + 6 =		43.	3 + 5 + 6 =	
22.	3 + 6 + 9 =		44.	2 + 6 + 9 =	

B

Number Correct: _____

Improvement: _____

Adding to the Teens

1.	5 + 5 + 4 =	
2.	7 + 3 + 5 =	
3.	1 + 9 + 8 =	
4.	4 + 6 + 2 =	
5.	2 + 8 + 9 =	
6.	7 + 0 + 6 =	
7.	4 + 0 + 9 =	
8.	2 + 9 + 9 =	
9.	4 + 5 + 4 =	
10.	8 + 7 + 5 =	
11.	2 + 7 + 9 =	
12.	9 + 1 + 2 =	
13.	6 + 4 + 5 =	
14.	8 + 2 + 3 =	
15.	1 + 4 + 9 =	
16.	3 + 8 + 0 =	
17.	7 + 4 + 7 =	
18.	5 + 3 + 8 =	
19.	7 + 3 + 4 =	
20.	5 + 8 + 6 =	
21.	7 + 6 + 4 =	
22.	5 + 8 + 4 =	

23.	8 + 2 + 5 =	
24.	9 + 1 + 6 =	
25.	3 + 6 + 4 =	
26.	3 + 2 + 7 =	
27.	4 + 8 + 6 =	
28.	9 + 9 + 0 =	
29.	0 + 7 + 5 =	
30.	8 + 4 + 4 =	
31.	3 + 8 + 8 =	
32.	5 + 7 + 6 =	
33.	3 + 4 + 9 =	
34.	3 + 7 + 3 =	
35.	6 + 4 + 5 =	
36.	7 + 9 + 1 =	
37.	2 + 6 + 8 =	
38.	5 + 3 + 7 =	
39.	6 + 0 + 9 =	
40.	2 + 5 + 7 =	
41.	3 + 6 + 3 =	
42.	4 + 2 + 9 =	
43.	6 + 3 + 5 =	
44.	7 + 2 + 9 =	

Lesson 4:   Count up to 1,000 on the place value chart.

EUREKA
MATH™

Name _____   Date _____

Work with your partner. Imagine your place value chart. Write down how you might count from the first number up to the second number. Underline the numbers where you bundled to make a larger unit.

1. 476 to 600

2. 47 to 200

3. 188 to 510

4. 389 to 801

©2015 Great Minds. eureka-math.org
G2-M3-TE-B2-1.3.1-01.2016

Name _____   Date _____

1. These are bundles of 10. If you put them together, which unit will you make?

   a. one   b. ten   c. hundred   d. thousand

2. These are bundles of hundreds, tens, and ones. How many sticks are there in all?

   _____

3. Imagine the place value chart. Write the numbers that show a way to count from 187 to 222.

EUREKA
MATH

Name _____   Date _____

1. Marcos used the place value chart to count bundles. How many sticks does Marcos have in all?

Hundreds	Tens	Ones

Marcos has _____ sticks.

2. Write the number:

Hundreds	Tens	Ones

3. These are hundreds. If you put them together, which unit will you make?

a. one   b. hundred   c. thousand   d. ten

4. Imagine 585 on the place value chart. How many ones, tens, and hundreds are in each place?

_____ _____ _____
       ones   tens   hundreds

5. Fill in the blanks to make a true number sentence.

   12 ones = _____ ten _____ ones

6. Show a way to count from 170 to 410 using tens and hundreds.
   Circle at least 1 benchmark number.

7. Mrs. Sullivan's students are collecting cans for recycling. Frederick collected 20 cans, Donielle collected 9 cans, and Mina and Charlie each collected 100 cans. How many cans did the students collect in all?

EUREKA
MATH™

©2015 Great Minds. eureka-math.org
G2-M3-TE-B2-1.3.1-01.2016

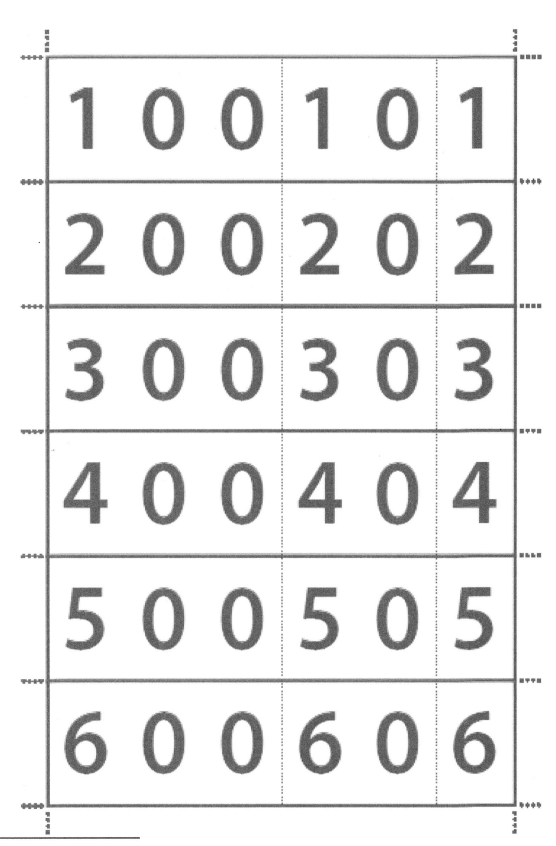

hide zero cards

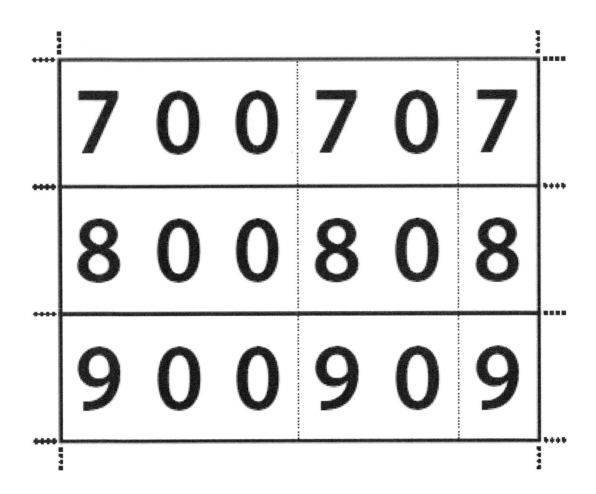

Lesson 4:   Count up to 1,000 on the place value chart.

EUREKA
MATH

ones	tens	hundreds

hundreds place value chart

©2015 Great Minds. eureka-math.org
G2-M3-TE-B2-1.3.1-01.2016

Lesson 5

Objective: Write base ten three-digit numbers in unit form; show the value of each digit.

Suggested Lesson Structure

■ Fluency Practice   (12 minutes)
▨ Application Problem   (10 minutes)
▢ Concept Development   (28 minutes)
▨ Student Debrief   (10 minutes)

Total Time   **(60 minutes)**

Fluency Practice (12 minutes)

- Exchange to Get to 100 **2.NBT.1a**   (5 minutes)
- Meter Strip Addition **2.NBT.5**   (7 minutes)

Exchange to Get to 100 (5 minutes)

Materials: (S) Dienes blocks: 12 ones, 10 tens, and 1 hundred per student; 1 die per pair

To keep student engagement high, you might modify the game for the class or for individuals. These are some adjustment suggestions:

- Two pairs at a table can "race" against each other rather than compete individually. This provides support and may reduce anxiety for students working below grade level or students with disabilities.
- Students working below grade level or those with disabilities may benefit from writing the new total after each turn.
- Switch the game to become Exchange to Get to 0. Students start at 100 and subtract the number of ones rolled on the die, exchanging tens rods for ones cubes.

Meter Strip Addition: Using Two-Digit Numbers with Totals in the Ones Place that Are Less Than or Equal to 12 (7 minutes)

Materials: (S) Meter strip (Lesson 1 Fluency Template)

    T: (Each student has a meter strip.) We're going to practice addition using our meter strips.

    T: Put your finger on 0. Slide up to 20. (Wait.) Slide up 9 more.

    T: How many centimeters did you slide up altogether?

    S: 29 centimeters.

    T: Tell your partner a number sentence describing sliding from 20 to 29.

Lesson 5:   Write base ten three-digit numbers in unit form; show the value of each digit.

©2015 Great Minds. eureka-math.org
G2-M3-TE-B2-1.3.1-01.2016

S: 20 + 9 = 29.

T: Put your finger on 0. Slide up to 34. (Wait.) Slide up 25 more.

T: How many centimeters did you slide up altogether?

S: 59 centimeters!

T: Whisper a number sentence describing sliding from 34 to 59.

S: 34 + 25 = 59

Continue with the following possible sequence: 46 + 32, 65 + 35, 57 + 23, 45 + 36, and 38 + 24.

Application Problem (10 minutes)

Freddy has $250 in ten-dollar bills.

a.   How many ten-dollar bills does Freddy have?

b.   He gave 6 ten-dollar bills to his brother. How many ten-dollar bills does he have left?

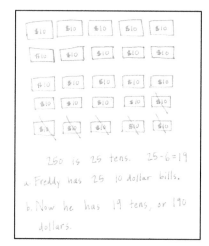

T: Let's read this problem together.

T: Talk with your partner about how you can draw the information given in the problem.

T: (Circulate. Listen for clear, concise explanations, as well as creative approaches to solving.)

S: I drew tens and skip-counted by 10 all the way up to 250. → I counted by tens up to $250 and kept track with a tally. → I skip-counted by tens to 100. That was 10 tens so then I just added 10 tens and then 5 tens. → I know 10 tens are in 100, so I drew 2 bundles of 100 and wrote 10 under each one. And I know 50 is 5 tens. So I counted 10, 20, 25 tens.

T: How many ten-dollar bills does Freddy have?

S: Freddy has 25 ten-dollar bills.

T: Please add that statement to your paper.

T: Now talk with your partner about Part B of this problem. Can you use your drawing to help you solve? (After a minute.)

S: I crossed off 6 tens and counted how many were left.

T: Raise your hand if you did the same thing? Who solved it another way?

S: I counted on from 6 tens.

T: I hear very good thinking! So tell me, how many ten-dollar bills does Freddy have left?

S: Freddy has 19 ten-dollar bills!

T: Add that statement to your paper.

NOTES ON
MULTIPLE MEANS
OF ENGAGEMENT:

Invite students to analyze different solution strategies. If you have the technical capability, project carefully selected student work two at a time. This is an argument for having word problems on half sheets of paper to facilitate comparison. Assign students the same problem for homework. This gives them the chance to try one of the new strategies.

Use the names of students in your class and culturally relevant situations within story problems to engage students. For example, Freddy is a student in this class.

EUREKA
MATH™

Lesson 5:   Write base ten three-digit numbers in unit form; show the value of
                 each digit.

©2015 Great Minds. eureka-math.org
G2-M3-TE-B2-1.3.1-01.2016

73

Concept Development (28 minutes)

Materials: (T) Bundles of straws from Lesson 1, place value box from Lesson 4, Hide Zero cards (Lesson 4
Template 1) (S) Hide Zero cards 1–5, 10–50, and 100–500 (Lesson 4 Template 1) cut apart
(as pictured) and in a small resealable bag

T: (Have 4 ones, 3 tens, and 2 hundreds already in the place value
box.) Count for me.

S: 1 one, 2 ones, 3 ones, 4 ones. 1 ten, 2 tens, 3 tens.
1 hundred, 2 hundreds.

T: Can I make larger units?

S: No!

T: In order from largest to smallest, how many of each unit are
there?

S: 2 hundreds, 3 tens, 4 ones.

T: What number does that represent?

S: 234.

T: What if we have 3 tens, 4 ones, and 2 hundreds. What number does
that represent?

S: 234.

T: (Show 234 with Hide Zero cards. Pull the cards apart to
show the value of each digit separately. Push them
back together to unify the values as one number.)
Open your bag. Build the number 234 with your Hide
Zero cards.

S: (Find the cards in their bags and build the number.)

T: Which of your cards shows this number of straws?
(Hold up 2 hundreds.) This number of straws? (Hold
up 4 ones.) Which has greater value, 2 hundreds or
4 ones?

S: 2 hundreds.

T: (Write on the board ____ hundreds ____ tens ____
ones.) Tell me the number of each unit. (Point to the
number modeled in the place value box.)

S: 2 hundreds 3 tens 4 ones.

T: That is called **unit form**.

T: We can also write this number as (write on board) two hundred thirty-four. This is the **word form**.

T: Work with your partner with your Hide Zero cards showing 234. Pull the cards apart and push them
together. Read the number in unit form and in word form.

**NOTES ON
MULTIPLE MEANS
OF ACTION AND
EXPRESSION:**

Remember, not all students will
complete the same amount of work.
Provide extra examples for early
finishers such as adding to the number
of ones, tens, and hundreds in the
place value boxes. Provide more
examples at a simpler level for students
who need additional practice before
moving on to numbers with zeros, such
as those in the Problem Set below.

Lesson 5:   Write base ten three-digit numbers in unit form; show the value of
each digit.

©2015 Great Minds. eureka-math.org
G2-M3-TE-B2-1.3.1-01.2016

Guide students through the following sequence of activities.

- Model numbers in the place value box.
- Students represent them with their Hide Zero cards.
- Students say the number in word form and unit form.

A suggested sequence might be: 351, 252, 114, 144, 444, 250, and 405. These examples include numbers that repeat a digit and those with zeros. Also, in most of the examples, the numbers have digits that are smaller in the hundreds place than in the tens or ones. While circulating, ask questions such as, "Which has more value, this 4 or this 4?" "What is the meaning of the zero?"

Problem Set (10 minutes)

Students should do their personal best to complete the Problem Set within the allotted 10 minutes. For some classes, it may be appropriate to modify the assignment by specifying which problems they work on first. Some problems do not specify a method for solving. Students should solve these problems using the RDW approach used for Application Problems.

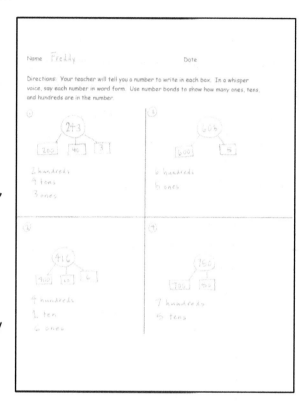

T: (Tell students the following numbers to model on the Problem Set: 243, 416, 605, and 750.)

Note: The Problem Set advances to numbers not within students' set of Hide Zero cards. Have students represent each number with number bonds where each part is shaped like a Hide Zero card. As needed, represent the numbers in the place value boxes. Hold all students accountable for saying each number in a whisper voice.

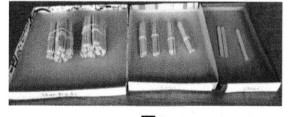

Lesson 5:   Write base ten three-digit numbers in unit form; show the value of each digit.

75

©2015 Great Minds. eureka-math.org
G2-M3-TE-B2-1.3.1-01.2016

Student Debrief (10 minutes)

Lesson Objective: Write base ten three-digit numbers in unit form; show the value of each digit.

Materials: (T) Blank paper to write numerals (as pictured on the previous page), place value box, bundles of straws for modeling (S) Individual place value charts (Template), personal white board

The Student Debrief is intended to invite reflection and active processing of the total lesson experience.

 - T: Bring your Problem Set to our Debrief. (Post, draw, or project a place value chart.)
 - T: Whisper this number to me. (Point to 243 on the Problem Set.)
 - S: 243.
 - T: (Model it with bundles in the place value box.) How many hundreds?
 - S: 2 hundreds.
 - T: (Replace the 2 hundreds with the digit 2.)
 - T: How many tens?
 - S: 4 tens.
 - T: (Replace the 4 tens with the digit 4.)
 - T: How many ones?
 - S: 3 ones.
 - T: (Replace the 3 ones with the digit 3.)
 - T: We now have represented 243 on the place value chart as a number. It is up to you to know the **MP.6** units represented and to remember that 2 hundreds has a different value than 2 ones.
 - T: (Write 416 and display in the place value box. Point to the number on the place value chart.) Say the value in **unit form**.
 - S: 4 hundreds 1 ten 6 ones.
 - T: (Point to the number 416.) Say the number in **word form**.
 - S: Four hundred sixteen.
 - T: (Write 605 and display in the place value box. Point to the number on the place value chart.) Say the value in unit form.
 - S: 6 hundreds 0 tens 5 ones.
 - T: (Point to the number 605.) Say the number in word form.
 - S: Six hundred five.
 - T: (Finish with 750.)

Students slide the individual place value charts template into their personal white boards. An example of a filled in template is pictured.

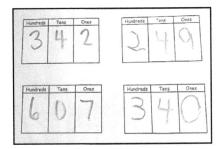

 - T: Turn to your partner. Partner A, write a number in your place value chart. Partner B says the number in unit form and then word form. Then, switch roles.

Lesson 5:   Write base ten three-digit numbers in unit form; show the value of each digit.

Circulate and listen for one minute. Take notes as necessary. Students should speak with confidence. Note those that still have insecurity even if their answers are correct. This is evidence that they simply need more practice.

T: (Display a set of four numbers as pictured.) What is the value of this 6? Answer in a complete sentence using the sentence frame: "The value is ___."

S: The value is 6 hundreds.

T: What is the value of this 6?

S: The value is 6 tens.

Hundreds	Tens	Ones
6	4	2

Hundreds	Tens	Ones
2	0	6

Hundreds	Tens	Ones
2	6	4

Hundreds	Tens	Ones
6	4	0

T: You knew the different values because you saw where I pointed. The place told you the value.

T: Tell your partner how you knew the value of each 6.

S: Because one was here and one was in the middle. → Because that is where we had bundles of tens and hundreds. → Because it says hundreds and tens there above. → Because this 6 was in the hundreds place and this 6 was in the tens place.

T: What is the first number on our chart?

S: 642.

T: Look, 264 has 2, 4, and 6 but in different places! The place tells us the value.

T: We call this a place value chart because each place (point to each place) has a value. We use 0–9 but their place tells us the unit represented.

T: Take turns telling your partner each of these numbers in unit form and in word form. If you finish early, write an interesting number for your partner to analyze. (Allow a few minutes.)

T: Let's close our lesson by having you explain to your partner what a place value chart is. Use the words *value, unit,* and *place.*

Exit Ticket (3 minutes)

After the Student Debrief, instruct students to complete the Exit Ticket. A review of their work will help with assessing students' understanding of the concepts that were presented in today's lesson and planning more effectively for future lessons. The questions may be read aloud to the students.

   Lesson 5:   Write base ten three-digit numbers in unit form; show the value of   77
                          each digit.

©2015 Great Minds. eureka-math.org
G2-M3-TE-B2-1.3.1-01.2016

Name _____   Date _____

Your teacher will tell you a number to write in each box. In a whisper voice, say each number in word form. Use number bonds to show how many ones, tens, and hundreds are in the number.

Lesson 5:   Write base ten three-digit numbers in unit form; show the value of each digit.

EUREKA MATH™

Name _____   Date _____

1. Look at the Hide Zero cards. What is the value of the 6?

a. 6   b. 600   c. 60

2. What is another way to write 5 ones 3 tens 2 hundreds?

a. 325   b. 523   c. 253   d. 235

3. What is another way to write 6 tens 1 hundred 8 ones?

a. 618   b. 168   c. 861   d. 681

4. Write 905 in unit form.

Lesson 5:   Write base ten three-digit numbers in unit form; show the value of
each digit.

©2015 Great Minds. eureka-math.org
G2-M3-TE-B2-1.3.1-01.2016

79

Name _____   Date _____

1. What is the value of the 7 in | 7 | 6 | 4 | ? _____

2. Make number bonds to show the hundreds, tens, and ones in each number. Then, write the number in unit form.

    a. 333

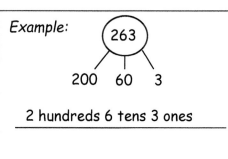

Example: 263

200   60   3

2 hundreds 6 tens 3 ones

_____

    b. 330

_____

    c. 303

Lesson 5:   Write base ten three-digit numbers in unit form; show the value of each digit.

EUREKA
MATH™

3. Draw a line to match unit form with number form.

   a. 1 hundred 1 one =   11

   b. 1 ten 1 one =   710

   c. 7 tens 1 one =   110

   d. 7 hundreds 1 one =   701

   e. 1 hundred 1 ten =   101

   f. 7 hundreds 1 ten =   71

hundreds	tens	ones

hundreds	tens	ones

hundreds	tens	ones

hundreds	tens	ones

individual place value charts

Lesson 5: Write base ten three-digit numbers in unit form; show the value of each digit.

EUREKA MATH™

©2015 Great Minds. eureka-math.org
G2-M3-TE-B2-1.3.1-01.2016

Lesson 6

Objective: Write base ten numbers in expanded form.

Suggested Lesson Structure

■ Fluency Practice   (12 minutes)
▨ Application Problem   (8 minutes)
▨ Concept Development   (30 minutes)
■ Student Debrief   (10 minutes)

Total Time   **(60 minutes)**

Fluency Practice (12 minutes)

- Meter Strip Addition **2.NBT.5**   (7 minutes)
- Unit Form Counting from 398 to 405 **2.NBT.3**   (3 minutes)
- Think 10 to Add 9 **2.OA.2**   (2 minutes)

Meter Strip Addition: With Two-Digit Numbers and Totals in the Ones that Are Greater Than 12 (7 minutes)

Materials: (S) Meter strip (Lesson 1 Fluency Template), personal white board

- T: (Each student has a meter strip.) We're going to practice addition using our meter strips.
- T: Put your finger on 0. Slide up to 27 centimeters. (Wait.) Slide up 35 more centimeters. You might first skip-count by ten three times, and then go up 5 ones.
- T: How many centimeters did you slide up altogether?
- S: 62 centimeters.
- T: Tell your partner a number sentence describing sliding from 27 to 62.
- S: 27 + 35 = 62.
- T: Put your finger on 0 centimeters. Slide up to 38 centimeters. (Wait.) Slide up 36 more centimeters.
- T: How many centimeters did you slide up altogether?
- S: 74 centimeters!
- T: At the signal, say a number sentence describing sliding from 38 to 74. (Signal.)
- S: 38 + 36 = 74.

Continue with the following possible sequence: 37 + 37, 45 + 28, 49 + 26, 68 + 28, and 57 + 29.

©2015 Great Minds. eureka-math.org
G2-M3-TE-B2-1.3.1-01.2016

T: In each of these problems we had more than 9 ones, so we had to make a new ten. I will write an expression. Wait for the signal. Say, "Make ten," if you have more than 9 ones. Say, "You can't make ten," if there are not enough ones.

T: 35 + 22.

S: You can't make ten.

T: 63 + 16.

S: You can't make ten.

T: 48 + 29.

S: Make ten.

T: 36 + 54.

S: Make ten.

T: 27 + 16.

S: Make ten.

T: Now, turn to your partner, and on your personal white board, write as many addition expressions as you can solve on your meter strip that need to make ten. You have one minute. Take your mark, get set, go!

Unit Form Counting from 398 to 405 (3 minutes)

Materials: (T) Hide Zero cards (Lesson 4 Template 1)

MP.6

T: Today we're going to practice unit form counting. This time we'll include hundreds! The unit form way to say 324 is *3 hundreds 2 tens 4 ones*. (Pull the cards apart to show the 300, 20, and 4.)

T: Try this number. (Show 398. Signal.)

S: 3 hundreds 9 tens 8 ones.

T: (Pull cards apart.) That's right!

T: Let's count on from 398 the unit form way. (Display 399–405 with Hide Zero cards as students count.)

S: 3 hundreds 9 tens 9 ones, 4 hundreds, 4 hundreds 1 one, 4 hundreds 2 ones, 4 hundreds 3 ones, 4 hundreds 4 ones, 4 hundreds 5 ones.

Think 10 to Add 9 (2 minutes)

T: Listen carefully! If I say, "9 + 5," you say, "10 + 4." Wait for my signal. Ready?

T: 9 + 5.

S: 10 + 4.

T: 9 + 3.

S: 10 + 2.

T: 9 + 7.

S: 10 + 6.

©2015 Great Minds. eureka-math.org
G2-M3-TE-B2-1.3.1-01.2016

T: 9 + 4.

S: 10 + 3.

T: 9 + 2.

S: 10 + 1.

T: 9 + 6.

S: 10 + 5.

T: 9 + 9.

S: 10 + 8.

T: 9 + 8.

S: 10 + 7.

Application Problem (8 minutes)

Timmy the monkey picked 46 bananas from the tree. When he was done, there were 50 bananas left. How many bananas were on the tree at first?

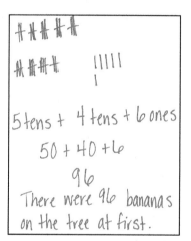

T: Read the problem with me.

T: Close your eyes and visualize Timmy the monkey and all those bananas.

T: Talk with your partner: What can you draw to show what you see?

S: I can draw the 46 bananas Timmy picked, and I can draw 50 bananas that are still on the tree.

T: What is the question asking? Read it again.

S: How many bananas were on the tree at first?

T: *At first* means at the very beginning of the story, before Timmy picked any bananas.

T: Work with your partner. How many different ways can you find the answer? (Circulate and listen for different strategies.)

T: Who would like to share their thinking?

S: At the beginning, all the bananas were on the tree. So I drew 4 tens 6 ones and 5 tens, and then I added and got 9 tens 6 ones, 96. → I know 50 is 5 tens, so I counted on 5 tens from 46: 56, 66, 76, 86, 96. → I made a number bond of 46 as 40 and 6, and then I wrote 50, and 40 plus 50 is 90, plus 6 more is 96.

T: Such creative problem solving! And did we all get the same answer?

S: Yes!

T: So how many bananas were on the tree at first? Give me a complete sentence.

S: 96 bananas were on the tree at first!

T: Yes! Please add that statement to your paper.

Lesson 6:   Write base ten numbers in expanded form.

©2015 Great Minds. eureka-math.org
G2-M3-TE-B2-1.3.1-01.2016

Concept Development (30 minutes)

Materials: (T) Place value box, bundles of straws for modeling (S) Hide Zero cards (Lesson 4 Template 1), math journal or paper

Expanded Form in Unit Order (8 minutes)

T: (Have the number 243 both written and modeled in the place value box.) Read this number to me in unit form. (Point.)

S: 2 hundreds 4 tens 3 ones.

T: Count for me up to 243 using the bundles in my place value box. (Record their count numerically by unit on the board in a single line horizontally as pictured to the right.)

100	100	10	10	10	10	1	1	1

S: 1 hundred, 2 hundred, 2 hundred ten, 2 hundred twenty, …

T: Each time we count a new unit, we are adding it to what we had before. Let's reread this putting in addition symbols.

T: (Write in the symbols as students read.)

S: $100 + 100 + 10 + 10 + 10 + 10 + 1 + 1 + 1 = 243$.

T: Explain to your partner why this is the same as 243.

$\underline{100 + 100} + \underline{10 + 10 + 10 + 10} + \underline{1 + 1 + 1} = 243$

$200 \qquad\quad + 40 \qquad\quad + 3 \quad = 243$

T: (Point to 100 + 100.) The answer is…?

S: 200.

T: (Write it below. Then point to 10 + 10 + 10 + 10.) The answer is…?

S: 40.

T: (Write it below. Then point to 1 + 1 + 1.) The answer is…?

S: 3.

T: (Write it below.) $100 + 100 + 10 + 10 + 10 + 10 + 1 + 1 + 1 = ?$

T: Say the number in unit form.

S: 2 hundreds 4 tens 3 ones.

T: (Point to the number sentence.) Are there 2 hundreds? 4 tens? 3 ones?

S: Yes!

T: 200 + 40 + 3 is…?

S: 243.

T: Show 243 with your Hide Zero cards.

T: 240 + 3 is…?

S: 243.

T: 200 + 43 is…?

S: 243.

T: There are different ways we can write our number. Now, let's add the total value of each unit.

Lesson 6:   Write base ten numbers in expanded form.

EUREKA
MATH™

©2015 Great Minds. eureka-math.org
G2-M3-TE-B2-1.3.1-01.2016

T: What is the total value of the hundreds in 243?

S: 200.

T: (Continue with the tens and ones.)

T: Now again, give me an addition sentence that adds up the total value of each of the units.

S: 200 + 40 + 3 = 243.

Problem Set Side 1 (7 minutes)

Students should do their personal best to complete the Problem Set within the allotted 7 minutes. For some classes, it may be appropriate to modify the assignment by specifying which problems they work on first. Some problems do not specify a method for solving. Students should solve these problems using the RDW approach used for Application Problems.

T: Excellent. Now you'll practice that so you get really good at it. On your Problem Set, write each number as an addition sentence separating the total value of each of the units. Let's do the first one together. (Guided practice.)

T: You have five minutes to do your personal best.

Name Freddy	Date
Write each number in expanded form, separating the total value of each of the units.	
1. 231 $200 + 30 + 1 = 231$	2. 312 $300 + 10 + 2 = 312$
3. 527 $500 + 20 + 7 = 527$	4. 752 $700 + 50 + 2 = 752$
5. 201 $200 + 1 = 201$	6. 310 $300 + 10 = 310$
7. 507 $500 + 7 = 507$	8. 750 $700 + 50 = 750$

Guided Practice: Expanded Form out of Unit Order (10 minutes)

T: Let's move the units around. (Pick up the 2 bundles of hundreds and move them to the end.)

T: (Write.) 10 + 10 + 10 + 10 + 1 + 1 + 1 + 100 + 100 = ?

T: Explain to your partner why this is the same as 243.

T: (Point to the number sentence.) Are there 2 hundreds? 4 tens? 3 ones?

S: Yes!

T: 40 + 3 + 200 is…?

S: 243.

$$\underline{10 + 10 + 10 + 10} + \underline{1 + 1 + 1} + \underline{100 + 100} = 243$$
$$40 \qquad\qquad + 3 \quad + 200 \quad = 243$$

T: Can someone explain what they understand about the order of the units and the total value? Talk about it with your partner first. (After a few moments.)

NOTES ON
MULTIPLE MEANS
OF REPRESENTATION:

When completing the Problem Set, let struggling students use the base ten materials, either the Dienes blocks or the straw bundles. The concrete representation often helps trigger the language and improves their confidence. Transition them to the abstract number and its word form by hiding the materials rather than taking them away entirely.

S: I notice that 3 and 40 and 200 is the same as 200 and 3 and 40. They're both 243. → We can write the units in any order, but the total stays the same. → It doesn't matter which unit we say first. It all adds up to the same amount.

T: Yes! It's important to be on the lookout for patterns and structures you can use to make sense out of math!

T: You've discovered there are different ways we can write our units, but the order does not affect the totals.

T: What is 2 + 4 + 3?

S: 9.

T: What is 3 + 4 + 2?

S: 9.

T: Explain to your partner why these totals are equal.

S: 6 + 3 is 9 and 7 + 2 is 9, too. → When you add, it doesn't matter if the parts are switched around. → You can make both problems 5 + 4 just by adding 2 and 3 first. → You can make both sides equal to 7 + 2 just by adding the 4 and 3 first.

T: Is the same true if our numbers are larger?

$$2 + 4 + 3 = 3 + 4 + 2$$

$$9 = 9$$

$$200 + 40 + 3 = 40 + 3 + 200$$

$$243 = 243$$

S: Yes!

Problem Set Side 2 (5 minutes)

T: Excellent. Let's practice that so you get really good at it. I have written some addition problems that tell the total value of each unit. Please write the total value in numerals. Be careful because they are not in order from the largest to the smallest unit!

Write the answer in number form.

9. 2 + 30 + 100 = 132	10. 300 + 2 + 10 = 312
11. 50 + 200 + 7 = 257	12. 70 + 500 + 2 = 572
13. 1 + 200 = 201	14. 100 + 3 = 103
15. 700 + 5 = 705	16. 7 + 500 = 507

Student Debrief (10 minutes)

Lesson Objective: Write base ten numbers in expanded form.

The Student Debrief is intended to invite reflection and active processing of the total lesson experience.

Invite students to review their solutions for the Problem Set. They should check work by comparing answers with a partner before going over answers as a class. Look for misconceptions or misunderstandings that can be addressed in the Debrief. Guide students in a conversation to debrief the Problem Set and process the lesson.

EUREKA
MATH™

T: Bring your Problem Set to our lesson Debrief. Check your answers for two minutes in groups of three.

S: (Check answers.)

T: Now, I'm going to read the answers. If you got it correct, whisper, "Yes."

T: Number 1, 200 + 30 + 1 = 231.

S: Yes! (Continue through both sides of the Problem Set at a lively pace.)

T: Work out any mistakes you made for one minute. Ask your group for help if you need it.

S: (Work together.)

T: I have written up pairs of problems that I want you to compare. How are they the same? How are they different?

1 and 2	9 and 10
3 and 4	11 and 12
1 and 5	1 and 9
2 and 6	2 and 10
3 and 7	7 and 16
4 and 8	Look for other connections, too.

NOTES ON MULTIPLE MEANS OF ACTION AND EXPRESSION:

Not all students will finish copying the rows in the given time frame. Early finishers can generate their own examples of expanded form and trade papers with a partner to check their work.

T: When we write our numbers as addition sentences with parts representing the total value of each unit, it is called **expanded form**. It helps us to see the value of each place.

T: Let's write an example in our math journal. You have two minutes to do your personal best. Write *Expanded Form* and then write the following examples to help you. Start by copying the entire first row.

T: (Write *Expanded Form* on the board for students to copy.)

Examples:

200 + 40 + 9 = 249	9 + 40 + 200 = 249
900 + 10 + 3 = 913	913 = 3 + 900 + 10
400 + 3 = 403	3 + 400 = 403
200 + 50 = 250	250 = 200 + 50

Exit Ticket (3 minutes)

After the Student Debrief, instruct students to complete the Exit Ticket. A review of their work will help with assessing students' understanding of the concepts that were presented in today's lesson and planning more effectively for future lessons. The questions may be read aloud to the students.

©2015 Great Minds. eureka-math.org
G2-M3-TE-B2-1.3.1-01.2016

Name _____ Date _____

Write each number in expanded form, separating the total value of each of the units.

1. 231

2. 312

3. 527

4. 752

5. 201

6. 310

7. 507

8. 750

Lesson 6: Write base ten numbers in expanded form.

EUREKA MATH

©2015 Great Minds. eureka-math.org
G2-M3-TE-B2-1.3.1-01.2016

Write the answer in number form.

9. 2 + 30 + 100 =

10. 300 + 2 + 10 =

11. 50 + 200 + 7 =

12. 70 + 500 + 2 =

13. 1 + 200 =

14. 100 + 3 =

15. 700 + 5 =

16. 7 + 500 =

Name _____   Date _____

1. Write in number form.

    a. 10 + 10 + 1 + 1 + 100 + 100 + 100 = _____

    b. 400 + 70 + 6 = _____

    c. _____ = 9 + 700 + 10

    d. _____ = 200 + 50

    e. 2 + 600 = _____

    f. 300 + 32 = _____

2. Write in expanded form.

    a. 974 = _____

    b. 435 = _____

    c. 35 = _____

    d. 310 = _____

    e. 703 = _____

Lesson 6:   Write base ten numbers in expanded form.

EUREKA
MATH™

Name _____   Date _____

1. Match the numerals with the number names.

   a. Two hundred thirty

   b. Forty

   c. Nine hundred sixty

   d. Four hundred seventy

   e. Eight hundred fifty

   f. Five hundred nineteen

   g. Four hundred seventeen

   h. Fourteen

   i. Nine hundred thirteen

   j. Eight hundred fifteen

   k. Five hundred ninety

   l. Two hundred thirteen

   m. Nine hundred sixteen

   ▪ 14

   ▪ 913

   ▪ 470

   ▪ 916

   ▪ 519

   ▪ 815

   ▪ 213

   ▪ 40

   ▪ 230

   ▪ 960

   ▪ 417

   ▪ 850

   ▪ 590

2. Write the answer in number form.

    a. 1 + 1 + 1 + 1 + 10 + 10 + 10 + 10 + 100 + 100 = _____

    b. 300 + 90 + 9 = _____

    c. _____ = 5 + 100 + 20

    d. _____ = 600 + 50

    e. 3 + 400 = _____

    f. 900 + 76 = _____

3. Write each number in expanded form.

    a. 533 = _____

    b. 355 = _____

    c. 67 = _____

    d. 460 = _____

    e. 801 = _____

Lesson 6:   Write base ten numbers in expanded form.

©2015 Great Minds. eureka-math.org
G2-M3-TE-B2-1.3.1-01.2016

EUREKA
MATH™

Lesson 7

Objective: Write, read, and relate base ten numbers in all forms.

Suggested Lesson Structure

■ Fluency Practice   (15 minutes)
▨ Concept Development   (27 minutes)
▨ Application Problem   (8 minutes)
■ Student Debrief   (10 minutes)
  Total Time   **(60 minutes)**

Fluency Practice (15 minutes)

- Place Value **2.NBT.1**, **2.NBT.3**   (4 minutes)
- Sprint: Expanded Form **2.NBT.3**   (8 minutes)
- Skip-Count Up and Down by $10 Between $45 and $125 **2.NBT.2**, **2.NBT.8**   (3 minutes)

Place Value (4 minutes)

Note: This fluency activity reviews place value concepts to prepare students for today's lesson.

T: (Write 157 on the board.) Say the number.
S: 157.
T: Say 157 in unit form.
S: 1 hundred 5 tens 7 ones.
T: Say 157 in expanded form.
S: 100 + 50 + 7.
T: What is 50 + 7 + 100?
S: 157.
T: What is 7 + 100 + 50?
S: 157.
T: How many ones are in 157?
S: 157 ones.
T: How many tens are in 157?
S: 15 tens.
T: What digit is in the ones place?
S: 7.

T: What is the value of the digit in the tens place?
S: 50.

Sprint: Expanded Form (8 minutes)

Materials: (S) Expanded Form Sprint

Skip-Count Up and Down by $10 Between $45 and $125 (3 minutes)

Materials: (T) 12 ten-dollar bills, 1 five-dollar bill

T: (Lay out $45 so that all students can see.)
    When I signal, tell the total value of the bills.
S: 45 dollars!
T: Good. Watch carefully as I change the total value.
    Count the new amount as I make it.
T: (Lay down ten-dollar bills to make $55, $65, $75, $85,
    $95, $105, $115, $125.)
S: (Respond in kind.)
T: (Take ten-dollar bills to make $115, $105, $95, $85,
    $75.)
S: (Respond in kind.)
T: (Lay down ten-dollar bills to make $85, $95, $105,
    $115, $125.)
S: (Respond in kind.)
T: (Take ten-dollar bills to make $115, $105, $95.)
S: (Respond in kind.)
T: (Continue alternating practice counting up and
    down, crossing back over numbers that students
    demonstrate difficulty counting.)

> **NOTES ON SCAFFOLDING FOR ELLs:**
>
> Sometimes a line of questioning can be further broken down to scaffold for English language learners or students working below grade level. It is much easier to see the change in the tens place when the value is expressed in unit form. Add the $10 and give wait time. "4 tens 5 dollars plus $10 is…? (provide sufficient wait time) 5 tens 5 dollars."

Concept Development (27 minutes)

Word Form, Unit Form, and Standard Form (5 minutes)

Materials: (S) Number spelling activity sheet (Activity
            Sheet), personal white board

T: From your spelling races (pictured to the right),
    I know you have worked hard to learn to read
    and spell numbers. You have two minutes to
    write as many numbers as you can.

EUREKA
MATH

S: (Write.)

T: Stop. Show me your personal white boards. (Review each one quickly.)

T: Check your work with your partner. Did you improve since the last time we did it? How many students spelled all the numbers to 10 correctly?

S: (Some students raise hands.)

T: Teen numbers. (Continue the process.)

T: Ok. I will show you a number in word or unit form. You write the number, we call that **standard form**, on your boards.

T: (Write *four hundred sixty-five*.)

S: (Write *465* and show at the signal.)

T: (Write *two hundred seventeen*.)

S: (Write *217* and show at the signal.)

T: (Write *nine hundred one*.)

S: (Write *901* and show at the signal.)

T: (Write *2 tens 7 hundreds 3 ones*.)

S: (Write *723* and show at the signal.)

T: (Write *13 tens 2 ones*.)

S: (Write *132* and show at the signal.)

Continue with more examples. Alternate between word form and unit form unless students need to focus on one type before moving on. We work toward mastery while avoiding predictable patterns that can lead to rigid thinking. The Problem Set provides a possible sequence to follow.

Problem Set Part 1 (5 minutes)

Students should do their personal best to complete the Problem Set within the allotted 5 minutes. For some classes, it may be appropriate to modify the assignment by specifying which problems they work on first. Some problems do not specify a method for solving. Students should solve these problems using the RDW approach used for Application Problems.

Materials: (S) Problem Set Part 1 (pictured to the right)

- A, B, C: Three non-zero digits in standard order.

- D, E, G, L: Word or unit form in standard order but with no ones or tens.

- H: Unit form with no tens.

- F, I, J, K: Unit form out of order.

- M: More than 9 units.

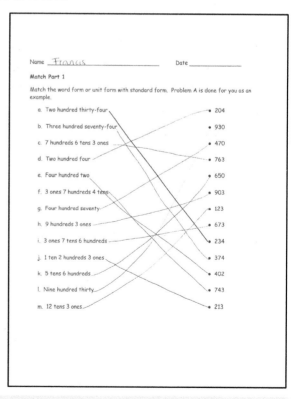

Lesson 7:   Write, read, and relate base ten numbers in all forms.   **97**

Students performing above grade level can be released to work on the Problem Set while keeping a group and giving them appropriate problems. Stay with a given type until they have it. Move on quickly when they do. Students who have completely caught on should move on to the Problem Set.

Early finishers might be provided with markers and sentence strips. They can create examples of the word or unit form to be used in the Student Debrief. The final problem is meant to lead into the next segment of the lesson.

More than 9 of a Unit (12 minutes)

Materials: (S) 21 ones and 21 tens per pair, personal white board

T: Partner A, with your straws, show me 12 ones.

S: (Count out 12 ones.)

T: Express 12 ones as tens and ones.

S: 1 ten 2 ones.

T: Tell the total value of 1 ten 2 ones.

S: 12.

T: Partner B, with your straws, show me 12 tens.

S: (Count out 12 tens.)

T: Express 12 tens as hundreds and tens.

S: 1 hundred 2 tens.

T: Tell the total value of 1 hundred 2 tens.

S: 120.

T: Turn to your partner and compare 12 ones to 12 tens.

S: They both have 12, but the units are different. → 12 ones makes 1 ten 2 ones, and 12 tens makes 1 hundred 2 tens. → 12 ones is 12, and 12 tens is 120.

T: Partner A, with your straws, show me 15 tens. Partner B, show me 15 ones.

T: Express to each other the value of your straws using the largest unit possible.

S: 1 ten 5 ones. → 1 hundred 5 tens.

T: Tell the total value of 1 hundred 5 tens.

S: 150.

T: Compare 15 ones to 15 tens with your partner.

S: (Compare.)

T: Partner A, show me 21 ones. Partner B, show me 21 tens.

T: Express 21 ones as tens and ones.

S: 2 tens 1 one.

T: Express 21 tens as hundreds and tens.

S: 2 hundreds 1 ten.

T: What is the total value of 2 hundreds 1 ten?

> **NOTES ON MULTIPLE MEANS OF ACTION AND EXPRESSION:**
>
> When a Problem Set involves significant reading, students need support from a visual. Post *hundred* with a picture of a bundle of one hundred, for example.
>
> Also, this is an opportunity for more fluent English language learners to show off and support their peers in small groups or partnerships.

S: 210.

T: Compare 21 ones to 21 tens with your partner.

S: (Compare.)

T: Put your straws away. On your personal white boards, express 68 ones as tens and ones.

S: (Write and show at the signal *6 tens 8 ones*.)

T: On your boards, express 68 tens as hundreds and tens.

S: 6 hundreds 8 tens.

T: Write the total value of 6 hundreds 8 tens.

S: (Write and show at the signal *680*.)

T: Compare 68 ones with 68 tens.

S: (Compare.)

Problem Set Part 2 (5 minutes)

Materials: (S) Problem Set Part 2 (pictured to the right)

Let those who are proficient work independently on the pictured Problem Set while a group stays for guided next questions. Notice that the examples of unit form now include addition. Let students make this connection. It should be well within their reach to make the jump.

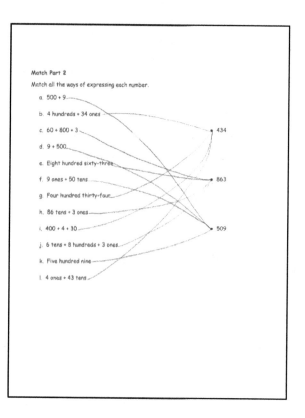

Application Problem (8 minutes)

Billy found a briefcase full of money. He counted up 23 ten-dollar bills, 2 hundred-dollar bills, and 4 one-dollar bills. How much money was in the briefcase?

T: Let's read this problem together.

T: Work with your partner to solve this problem. (Allow time for students to solve.) Who would like to share how they solved the problem?

S: I drew all the money, then I counted it. 100, 200, 210, 220, 230, 240, 250, …, 430, 431, 432, 433, 434. → I drew 23 circles to show 23 tens and counted up to 230 dollars. Then I skip-counted 200 more and got 430 dollars. Then I counted on 4 more dollars and got 434 dollars. → I added 200 + 4. That's just expanded form.

NOTES ON
PARENT CONFERENCES:

While circulating, make organized notes on what individual students are sharing. This informal assessment provides valuable feedback, which can be shared during parent conferences. As students realize that they are being held accountable even for these small dialogues, their performance improves. The message is that *we are listening* and find their ideas interesting.

S: Then I drew 23 tens and I skip-counted 2 hundreds and 3 tens from 204 and got 434. → I know 20 tens equals 200, so I counted on 2 more hundreds and got 400. Then I added the 3 tens from the 23 tens plus the 4 ones. 400 + 30 + 4 is 434. → I know 23 tens is 2 hundreds 3 tens. Add 2 more hundreds. That is 4 hundreds 3 tens, plus 4 ones makes 434. He had $434.

T: How many dollars were in the briefcase?

S: 434 dollars were in the briefcase.

T: Tell me the number in unit form.

S: 4 hundreds, 3 tens, 4 ones.

T: What is the number in expanded form?

S: 400 + 30 + 4.

T: Add the unit form, the expanded form, and the statement to your paper.

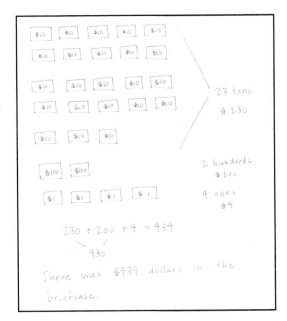

Student Debrief (10 minutes)

Lesson Objective: Write, read, and relate base ten numbers in all forms.

Materials: (S) Problem Set, Application Problem solution

The Student Debrief is intended to invite reflection and active processing of the total lesson experience.

Invite students to review their solutions for the Problem Set. They should check work by comparing answers with a partner before going over answers as a class. Look for misconceptions or misunderstandings that can be addressed in the Debrief. Guide students in a conversation to debrief the Problem Set and process the lesson.

T: Bring your Problem Set and the problem about Billy to our Debrief. Check your answers to the Problem Set with your partner for one minute. (Allow time for students to check.) I'll show the answers now. If you got the correct answer, say yes.

MP.6

T: Problem A 234 matches to (write 234).

S: Yes!

T: Problem B 374 matches to (write 374).

S: Yes!

T: (Move through the rest of the problems quickly.) Take one minute to correct your mistakes.

T: Would anyone like to share a mistake?

S: I got J wrong.

T: Why?

S: Because I saw 123 really close by and didn't read the words.

T: You didn't read the units?

S: Yes.

T: Someone else?

S: I got F wrong for the same reason.

T: What reason is that?

S: I didn't read the units so I chose 374.

T: That's why it's so important to understand the units when we read numbers in **standard form**. We have to be precise or we make mistakes. The Mars Climate Orbiter disintegrated in 1999 due to the use of the wrong units. NASA lost millions of dollars! Always be precise about your units.

T: Now, I want you to compare Problem Set Part 2 with the problem about Billy's money. Share with a partner.

MP.6

S: Billy has money, but this problem is just hundreds, tens, and ones. → Billy's ten-dollar bills are like the tens. → The problems all have enough tens to make a hundred. → Yeah, Billy has 23 ten-dollar bills. That means we can make 2 hundreds just like the problems that have 43 tens and 12 tens, so we can make a hundred. → 12 tens is 1 hundred 2 tens just like 23 ten-dollar bills is 2 hundred-dollar bills and 3 ten-dollar bills. All the problems have more than 9 units of ten. That means we can make groups of 10 tens or hundreds.

T: So, when we write numbers in unit form, sometimes there are more than 9 of a unit. That means we can make a larger unit!

S: Yes!

T: Excellent.

Exit Ticket (3 minutes)

After the Student Debrief, instruct students to complete the Exit Ticket. A review of their work will help with assessing students' understanding of the concepts that were presented in today's lesson and planning more effectively for future lessons. The questions may be read aloud to the students.

©2015 Great Minds. eureka-math.org
G2-M3-TE-B2-1.3.1-01.2016

Name _____   Date _____

Spell Numbers: How many can you write correctly in 2 minutes?

1		11		10	
2		12		20	
3		13		30	
4		14		40	
5		15		50	
6		16		60	
7		17		70	
8		18		80	
9		19		90	
10		20		100	

_____

number spelling activity sheet

Lesson 7:   Write, read, and relate base ten numbers in all forms.

©2015 Great Minds. eureka-math.org
G2-M3-TE-B2-1.3.1-01.2016

EUREKA
MATH™

A

Number Correct: _____

Expanded Form

1.	20 + 1 =		23.	400 + 20 + 5 =	
2.	20 + 2 =		24.	200 + 60 + 1 =	
3.	20 + 3 =		25.	200 + 1 =	
4.	20 + 9 =		26.	300 + 1 =	
5.	30 + 9 =		27.	400 + 1 =	
6.	40 + 9 =		28.	500 + 1 =	
7.	80 + 9 =		29.	700 + 1 =	
8.	40 + 4 =		30.	300 + 50 + 2 =	
9.	50 + 5 =		31.	300 + 2 =	
10.	10 + 7 =		32.	100 + 10 + 7 =	
11.	20 + 5 =		33.	100 + 7 =	
12.	200 + 30 =		34.	700 + 10 + 5 =	
13.	300 + 40 =		35.	700 + 5 =	
14.	400 + 50 =		36.	300 + 40 + 7 =	
15.	500 + 60 =		37.	300 + 7 =	
16.	600 + 70 =		38.	500 + 30 + 2 =	
17.	700 + 80 =		39.	500 + 2 =	
18.	200 + 30 + 5 =		40.	2 + 500 =	
19.	300 + 40 + 5 =		41.	2 + 600 =	
20.	400 + 50 + 6 =		42.	2 + 40 + 600 =	
21.	500 + 60 + 7 =		43.	3 + 10 + 700 =	
22.	600 + 70 + 8 =		44.	8 + 30 + 700 =	

Lesson 7: Write, read, and relate base ten numbers in all forms.

103

©2015 Great Minds. eureka-math.org
G2-M3-TE-B2-1.3.1-01.2016

B

Number Correct: _____

Improvement: _____

Expanded Form

1.	10 + 1 =	
2.	10 + 2 =	
3.	10 + 3 =	
4.	10 + 9 =	
5.	20 + 9 =	
6.	30 + 9 =	
7.	70 + 9 =	
8.	30 + 3 =	
9.	40 + 4 =	
10.	80 + 7 =	
11.	90 + 5 =	
12.	100 + 20 =	
13.	200 + 30 =	
14.	300 + 40 =	
15.	400 + 50 =	
16.	500 + 60 =	
17.	600 + 70 =	
18.	300 + 40 + 5 =	
19.	400 + 50 + 6 =	
20.	500 + 60 + 7 =	
21.	600 + 70 + 8 =	
22.	700 + 80 + 9 =	

23.	500 + 30 + 6 =	
24.	300 + 70 + 1 =	
25.	300 + 1 =	
26.	400 + 1 =	
27.	500 + 1 =	
28.	600 + 1 =	
29.	900 + 1 =	
30.	400 + 60 + 3 =	
31.	400 + 3 =	
32.	100 + 10 + 5 =	
33.	100 + 5 =	
34.	800 + 10 + 5 =	
35.	800 + 5 =	
36.	200 + 30 + 7 =	
37.	200 + 7 =	
38.	600 + 40 + 2 =	
39.	600 + 2 =	
40.	2 + 600 =	
41.	3 + 600 =	
42.	3 + 40 + 600 =	
43.	5 + 10 + 800 =	
44.	9 + 20 + 700 =	

Lesson 7:   Write, read, and relate base ten numbers in all forms.

EUREKA MATH™

Name _____   Date _____

Match Part 1

Match the word form or unit form with standard form. Problem A is done for you as an example.

a. Two hundred thirty-four   • 204

b. Three hundred seventy-four   • 930

c. 7 hundreds 6 tens 3 ones   • 470

d. Two hundred four   • 763

e. Four hundred two   • 650

f. 3 ones 7 hundreds 4 tens   • 903

g. Four hundred seventy   • 123

h. 9 hundreds 3 ones   • 673

i. 3 ones 7 tens 6 hundreds   • 234

j. 1 ten 2 hundreds 3 ones   • 374

k. 5 tens 6 hundreds   • 402

l. Nine hundred thirty   • 743

m. 12 tens 3 ones   • 213

Match Part 2

Match all the ways of expressing each number.

a. 500 + 9

b. 4 hundreds + 34 ones

c. 60 + 800 + 3   • 434

d. 9 + 500

e. Eight hundred sixty-three

f. 9 ones + 50 tens   • 863

g. Four hundred thirty-four

h. 86 tens + 3 ones

i. 400 + 4 + 30   • 509

j. 6 tens + 8 hundreds + 3 ones

k. Five hundred nine

l. 4 ones + 43 tens

Lesson 7: Write, read, and relate base ten numbers in all forms.

EUREKA MATH

©2015 Great Minds. eureka-math.org
G2-M3-TE-B2-1.3.1-01.2016

Name _____ Date _____

1. Write 342 in word form.

   _____

2. Write in standard form.

   a. Two hundred twenty-six _____

   b. Eight hundred three _____

   c. 5 hundreds + 56 ones _____

   d. 60 + 800 + 3 _____

3. Write the value of 17 tens three different ways. Use the largest unit possible.

   a. Standard form _____

   b. Expanded form _____

   c. Unit form _____

Lesson 7:   Write, read, and relate base ten numbers in all forms.

107

©2015 Great Minds. eureka-math.org
G2-M3-TE-B2-1.3.1-01.2016

Name _____ Date _____

These are bundles of hundreds, tens, and ones. Write the standard form, expanded form, and word form for each number shown.

1.

   a. Standard Form _____

   b. Expanded Form _____

   c. Word Form _____

2.

   a. Standard Form _____

   b. Expanded Form _____

   c. Word Form _____

EUREKA
MATH™

3. What is the unit value of the 3 in 432? _____

4. What is the unit value of the 6 in 216? _____

5. Write 212, 221, 122 in order from greatest to least.

   _____   _____   _____

Lesson 7:   Write, read, and relate base ten numbers in all forms.

109

©2015 Great Minds. eureka-math.org
G2-M3-TE-B2-1.3.1-01.2016

2
GRADE

Mathematics Curriculum

Topic D

Modeling Base Ten Numbers Within 1,000 with Money

2.NBT.2, 2.NBT.1, 2.NBT.3, 2.MD.8

Focus Standard:	2.NBT.2	Count within 1000; skip-count by 5s, 10s, and 100s.
Instructional Days:	3	
Coherence -Links from:	G1–M6	Place Value, Comparison, Addition and Subtraction to 100
-Links to:	G2–M4	Addition and Subtraction Within 200 with Word Problems to 100
	G2–M7	Problem Solving with Length, Money, and Data

Further building their place value understanding, students count by one dollar bills up to $124, repeating the process done in Lesson 4 with bundles. Using bills, however, presents a new option. A set of 10 ten-dollar bills can be traded or changed for 1 hundred-dollar bill, driving home the equivalence of the two amounts, an absolutely essential Grade 2 place value understanding (**2.NBT.1a**).

Next, students see that 10 bills can have a value of $10, $100, or $1,000 but appear identical aside from their printed labels (**2.NBT.1, 2.NBT.3**). A bill's value is determined by what it represents. Students count by ones, tens, and hundreds (**2.NBT.2**) to figure out the values of different sets of bills.

As students move back and forth from money to numerals, they make connections to place value that help them see the correlations between base ten numerals and corresponding equivalent denominations of one, ten, and hundred-dollar bills.

Word problems can be solved using both counting and place value strategies. For example, "Stacey has $154. She has 14 one-dollar bills. The rest is in $10 bills. How many $10 bills does Stacey have?" (**2.NBT.2**). Lesson 10 is an exploration to uncover the number of $10 bills in a $1,000 bill discovered in grandfather's trunk in the attic. (Note that the 1,000 dollar bill is no longer in circulation.)

A Teaching Sequence Toward Mastery of Modeling Base Ten Numbers Within 1,000 with Money

Objective 1: Count the total value of $1, $10, and $100 bills up to $1,000.
(Lesson 8)

Objective 2: Count from $10 to $1,000 on the place value chart and the empty number line.
(Lesson 9)

Objective 3: Explore $1,000. How many $10 bills can we change for a thousand dollar bill?
(Lesson 10)

Lesson 8

Objective: Count the total value of $1, $10, and $100 bills up to $1,000.

Suggested Lesson Structure

◼ Fluency Practice   (8 minutes)
◼ Application Problem   (8 minutes)
▢ Concept Development   (34 minutes)
◼ Student Debrief   (10 minutes)

Total Time   **(60 minutes)**

Fluency Practice (8 minutes)

- Mixed Counting with Ones, Tens, and Hundreds from 1,000 to 0 **2.NBT.2**   (5 minutes)
- Doubles **2.OA.2**   (1 minute)
- Related Facts Within 20 **2.OA.2**   (2 minutes)

Mixed Counting with Ones, Tens, and Hundreds from 1,000 to 0 (5 minutes)

Materials: (T) Bundle of one hundred, one ten, and a single stick from Lesson 1

- T: Let's play Mixed Counting using what we know about counting by ones, tens, and hundreds. I'll hold bundles to show you what to count by. A bundle of 100 means count by hundreds, a bundle of 10 means count by tens, and a single stick means count by ones.
- T: Let's start at 1,000 and count down. Ready? (Hold up a bundle of 10 until students count to 940. If necessary, create visual support with the difficult language of these numbers by writing them on the board as students count.)
- S: 990, 980, 970, 960, 950, 940.
- T: (Hold up a bundle of 100 until students count to 540.)
- S: 840, 740, 640, 540.
- T: (Hold up a bundle of 10 until students count to 500.)
- S: 530, 520, 510, 500.
- T: (Hold up a single one until students count to 495.)
- S: 499, 498, 497, 496, 495.
- T: (Hold up a ten until students count to 465.)
- S: 485, 475, 465.

Continue, varying practice counting with ones, tens, and hundreds down to zero.

EUREKA
MATH™

Doubles (1 minute)

T: I'll say a doubles fact. You tell me the answer. Wait for my signal. Ready?

T: 5 + 5.

S: 10.

T: 3 + 3.

S: 6.

T: 6 + 6.

S: 12.

T: 1 + 1.

S: 2.

T: 4 + 4.

S: 8.

T: 9 + 9.

S: 18.

T: 2 + 2.

S: 4.

T: 10 + 10.

S: 20.

T: 8 + 8.

S: 16.

T: 7 + 7.

S: 14.

Related Facts Within 20 (2 minutes)

T: I say, "10 – 6." You say, "6 + 4 = 10." Wait for my signal. Ready?

T: 8 – 3.

S: 3 + 5 = 8.

T: 13 – 7.

S: 7 + 6 = 13.

T: 11 – 8.

S: 8 + 3 = 11.

T: 15 – 9.

S: 9 + 6 = 15.

Continue in this manner for two minutes.

©2015 Great Minds. eureka-math.org
G2-M3-TE-B2-1.3.1-01.2016

Application Problem (8 minutes)

Stacey has $154. She has 14 one-dollar bills. The rest is in ten-dollar bills. How many ten-dollar bills does she have?

NOTES ON
MULTIPLE MEANS
OF ACTION AND
EXPRESSION:

T: Let's read this problem together.

T: Think for a moment, and then discuss with your partner: How does this problem relate to what we've been studying over the past several lessons? What similarities do you notice?

S: Money comes in tens and ones, too. → We've been learning about hundreds, tens, and ones, and money is just like that. → A ten-dollar bill is like a bundle of ten. → It's units of a hundred, ten, and one, just like with the straws. → It's like the place value chart but with money instead of numbers.

MP.2

T: How can making this connection help you solve the problem? Talk it over with your partner, and use what you've learned to solve. (Circulate and listen for discussions that rely on unit form, expanded form, and exchanging units to solve.)

S: I know 154 is 1 hundred 5 tens 4 ones. Stacey has 14 ones, and that's the same as 1 ten 4 ones. So, she needs 10 tens to make the hundred and 4 more tens to make 5 tens. She already has 4 ones. 10 tens plus 4 tens is 14 tens.

T: Outstanding reasoning, Valeria!

"Pretend Partner A is the parent, and Partner B is the child. Partner B, explain to your parent in your own words what Valeria just shared with the class."

This type of exchange gives students the opportunity to process key information. The language and thinking of the child who makes the original statement provides support for others. The process of reformulating the idea helps solidify understanding, and verbalizing it helps students clarify and internalize it.

T: Pretend Partner A is the parent and Partner B is the child. Partner B, explain to your parent in your own words what Valeria just shared with the class. Use words, numbers, and pictures to help your parent understand. Then, switch roles. (Allow students a few minutes.)

T: How many ten-dollar bills does Stacey have?

S: 140. → 14 ten-dollar bills.

T: I like the way many of you said the unit as part of your answer. It helps us be clear about whether we're answering the question correctly.

T: Reread the question.

S: How many ten-dollar bills does she have?

T: Does Stacey have 140 ten-dollar bills?

S: No.

T: Always check to be sure your answer makes sense. That's why it's important to answer the question with a statement. The question is not *how much money* does she have. It's *how many ten-dollar bills* does she have.

T: So, how many ten-dollar bills does Stacey have? Give me a complete sentence.

S: Stacey has 14 ten-dollar bills.

T: Good! Please add that statement to your paper.

Lesson 8:   Count the total value of $1, $10, and $100 bills up to $1,000.

©2015 Great Minds. eureka-math.org
G2-M3-TE-B2-1.3.1-01.2016

Concept Development (34 minutes)

Materials: (S) Personal white board, unlabeled hundreds place value chart (Template), 10 one-dollar bills, 10 ten-dollar bills, and 10 hundred-dollar bills (put money in a small resealable bag "wallet" with the ones in the front, tens in the middle, and hundreds in the back)

Part A: Counting by 1 Dollar up to $124 (6 minutes)

Note: Have students slide the place value chart template inside their personal white boards. Guide them to place 10 bills of each denomination above each column (pictured). Explain that the place value chart is unlabeled because the value is shown on each bill. Students encounter this again when they work with place value disks in Topic E.

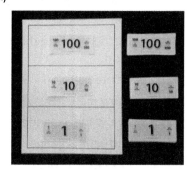

Directions:

1. Count up to $124 by one-dollar bills on your place value chart.
2. When you get 10 one-dollar bills, change them for 1 ten-dollar bill.
3. When you have 10 ten-dollar bills, change them for 1 hundred-dollar bill.
4. Whisper count the value of your money as you go.
5. Each time you make a change, let the other partner handle the money.

T: How is counting up to $124 with money bills different from counting up to 124 with bundles?

S: With straws, we could just get a rubber band. → With straws we bundled, but with money we changed 10 ones for 1 ten. → Yeah, we got a different bill from our wallet. → The 10 ten-dollar bills got changed for 1 hundred-dollar bill. → It was a trade, 10 things for 1 thing. → The hundred-dollar bill has a greater value, but it doesn't show.

Part B: Manipulating the Value of 10 Bills (6 minutes)

T: Partner A, put 5 one-dollar bills in a row horizontally across your desk.

T: Partner B, express the value of the money using this sentence frame. "The value of ____ dollar bills is ____."

S: The value of 5 one-dollar bills is $5.

T: Partner B, put another row of 5 one-dollar bills directly below the first row.

T: Partner A, express the new value of the money.

S: The value of 10 one-dollar bills is $10.

T: Partner A, change the 2 one-dollar bills on the far left for 2 ten-dollar bills.

T: Partner B, express the value of the money.

S: The value of 2 ten-dollar bills and 8 one-dollar bills is $28.

T: Partner B, write the value of the money in expanded form on your personal white board.

S: (Write $20 + $8 = $28.)

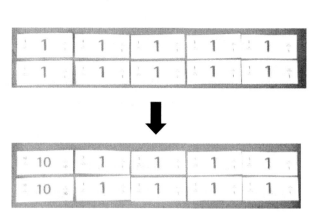

©2015 Great Minds. eureka-math.org
G2-M3-TE-B2-1.3.1-01.2016

T: Show me.

S: (Show the expanded form.)

T: Partner B, change the next 2 one-dollar bills on the left for 2 ten-dollar bills.

T: Partner A, express the value of the money.

S: The value of 4 ten-dollar bills and 6 one-dollar bills is $46.

T: Partner A, write the value of the money in expanded form.

S: (Write $40 + $6 = $46.)

T: Show me.

S: (Show the expanded form.)

T: Partner A, change 6 one-dollar bills to 6 ten-dollar bills.

T: Partner B, express the value of the money.

S: The value of 10 ten-dollar bills is $100.

> **NOTES ON MULTIPLE MEANS OF ENGAGEMENT:**
>
> Speaking and writing simultaneously is a powerful combination, giving students multi-modal input: oral, auditory, and kinesthetic. Whispering adds mystery and is therefore engaging. Circulate and listen intently to the math content in the students' speech. Encourage partners to listen intently, too.

Part C: Hundred-, Ten-, and One-Dollar Bills (11 minutes)

T: Show me $64.

T: Partner A, change the 2 ten-dollar bills on the left to 2 hundred-dollar bills.

T: What is the new value?

S: The value of 2 hundred-dollar bills, 4 ten-dollar bills, and 4 one-dollar bills is $244.

T: Write the value of the money in expanded form.

S: (Write $200 + $40 + $4 = $244.)

T: Show me.

S: (Show the expanded form.)

T: Partner B, change 1 ten-dollar bill on the left to 1 hundred-dollar bill.

T: What is the new value?

S: The value of 3 hundred-dollar bills, 3 ten-dollar bills, and 4 one-dollar bills is $334.

T: Write the value of the money in expanded form.

S: (Write $300 + $30 + $4 = $334.)

T: Show me.

S: (Show the expanded form.)

Continue as above using the following sequence:

- From $334, change 3 tens to 3 hundreds. (The new amount is $604.)
- From $604, change 4 ones to 4 tens. (The new amount is $640.)
- From $640, change 2 tens to 2 hundreds. (The new amount is $820.)
- From $820, change 1 ten to 1 one. (The new amount is $811.)
- From $811, change 1 ten to 1 one. (The new amount is $802.)
- From $802, change 2 ones to 2 hundreds. (The new amount is $1,000.)

EUREKA
MATH™

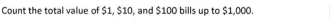

Problem Set (11 minutes)

Students should do their personal best to complete the Problem Set within the allotted 11 minutes. For some classes, it may be appropriate to modify the assignment by specifying which problems they work on first. Some problems do not specify a method for solving. Students should solve these problems using the RDW approach used for Application Problems.

Directions:

1. Represent each amount of money using 10 bills.
2. Write and whisper each amount of money in expanded form.
3. Write the total value of each set of bills as a number bond.

Student Debrief (10 minutes)

Lesson Objective: Count the total value of $1, $10, and $100 bills up to $1,000.

Materials: (T) 1 bundle of 100 straws
             (S) Completed Problem Set

The Student Debrief is intended to invite reflection and active processing of the total lesson experience.

Invite students to review their solutions for the Problem Set. They should check work by comparing answers with a partner before going over answers as a class. Look for misconceptions or misunderstandings that can be addressed in the Debrief. Guide students in a conversation to debrief the Problem Set and process the lesson.

- T: Discuss with your partner: Using any combination of $1, $10, and $100 bills, what is the smallest amount of money you can show with 10 bills, and what is the greatest amount of money you can show with 10 bills?
- T: (As students discuss the question, circulate and listen.)
- T: I heard many of you saying the smallest amount is…?

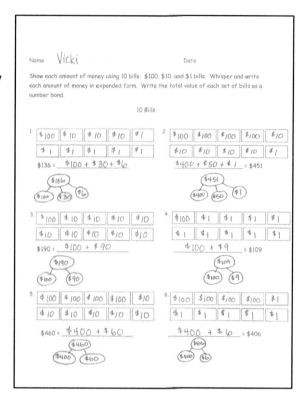

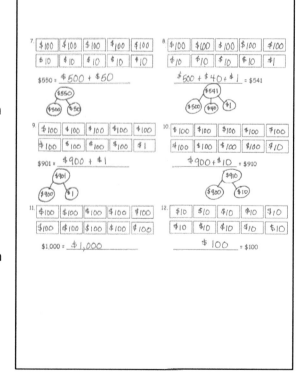

EUREKA MATH™

Lesson 8:   Count the total value of $1, $10, and $100 bills up to $1,000.

117

S: $10.

T: The greatest amount is…?

S: $1,000.

T: So, the value of the money changes, but what stays the same?

S: The size of the bills stays the same.

T: How do you know the value of the money?

S: By looking at it. → The value of each bill is written on it.

T: If you were blind, could you know its value?

S: No!

T: That's true here in the United States, but it's interesting to note that in other countries bills come in different sizes and even colors!

T: (Hold up a hundred bundle.) What about the value of this bundle? If you were blind, would you know? Talk to your partner about that.

S: Yes, because you could feel it was big. → Yes, because you could count the sticks. → Yes, because you could count the number of tens.

T: (Hold up a hundred-dollar bill.) Somebody decided this bill had a value of $100. But this bundle is 100 because it has 100 sticks, and we can count them.

T: Share your Problem Set with your partner. Compare answers and drawings for one minute.

T: I will read the answers now. If you got it correct, say "yes."

T: (Read the answers as students correct.)

T: Take a moment to analyze and talk about Problems 3 and 4: $190 and $109. What is different about the numbers?

S: $190 has no one-dollar bills and 9 ten-dollar bills. → $109 is less than $190 because it has 9 ones and no tens. → Wow. That is a big difference. Hmmm, that's 10, 20, 30, 40, 50, 60, 70, 80, 90. That's $81 more!

T: Do the same thing with Problems 2, 5, and 6: $451, $460, and $406. What is different about the numbers?

S: $460 is 9 dollars more than $451. → $460 and $406 switched the number of tens and ones. There are 6 ten-dollar bills in $460, but only 6 one-dollar bills in $406.

T: When you counted to $124, what happened when you had 10 one-dollar bills?

S: You could change them for 1 ten-dollar bill.

T: What happened when you had 10 ten-dollar bills?

S: You could change them for 1 hundred-dollar bill.

T: Which has a greater value, 3 hundred-dollar bills or 9 ten-dollar bills?

S: 3 hundred-dollar bills!

T: We counted the total value of many different amounts of money!

©2015 Great Minds. eureka-math.org
G2-M3-TE-B2-1.3.1-01.2016

Exit Ticket (3 minutes)

After the Student Debrief, instruct students to complete the Exit Ticket. A review of their work will help with assessing students' understanding of the concepts that were presented in today's lesson and planning more effectively for future lessons. The questions may be read aloud to the students.

Lesson 8:   Count the total value of $1, $10, and $100 bills up to $1,000.

119

©2015 Great Minds. eureka-math.org
G2-M3-TE-B2-1.3.1-01.2016

Name _____ Date _____

Show each amount of money using 10 bills: $100, $10, and $1 bills. Whisper and write each amount of money in expanded form. Write the total value of each set of bills as a number bond.

10 Bills

1.

$136 = _____

2.

_____ = $451

3.

$190 = _____

4.

_____ = $109

Lesson 8:   Count the total value of $1, $10, and $100 bills up to $1,000.

EUREKA
MATH™

5.

$460 = _____

6.

_____ = $406

7.

$550 = _____

8.

_____ = $541

EUREKA MATH

Lesson 8: Count the total value of $1, $10, and $100 bills up to $1,000.

121

©2015 Great Minds. eureka-math.org
G2-M3-TE-B2-1.3.1-01.2016

9.

$901 = _____

10.

_____ = $910

11.

$1,000 = _____

12.

_____ = $100

Lesson 8:   Count the total value of $1, $10, and $100 bills up to $1,000.

EUREKA MATH

Name _____   Date _____

1. Write the total value of the money shown below in standard and expanded form.

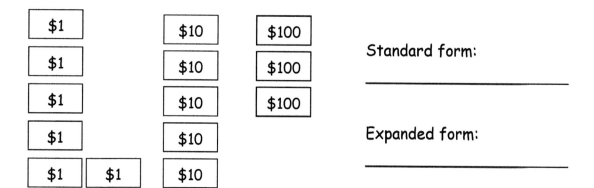

Standard form:

_____

Expanded form:

_____

2. What is the value of 3 ten-dollar bills and 9 one-dollar bills? _____

3. Draw money to show 2 different ways to make $142, using only $1, $10, and $100 bills.

EUREKA
MATH™

Lesson 8:   Count the total value of $1, $10, and $100 bills up to $1,000.

123

©2015 Great Minds. eureka-math.org
G2-M3-TE-B2-1.3.1-01.2016

Name _____   Date _____

1. Write the total value of the money.

$10	$10	$10	$10	$10
$10	$10	$10	$10	$1

_____

- -

$100	$100	$10	$1	$1
$1	$1	$1	$1	$1

_____

2. Fill in the bills with $100, $10, or $1 to show the amount.

$172

- -

$226

Lesson 8:   Count the total value of $1, $10, and $100 bills up to $1,000.

EUREKA MATH

3. Draw and solve.

Brandon has 7 ten-dollar bills and 8 one-dollar bills. Joshua has 3 fewer ten-dollar bills and 4 fewer one-dollar bills than Brandon. What is the value of Joshua's money?

Lesson 8:   Count the total value of $1, $10, and $100 bills up to $1,000.

©2015 Great Minds. eureka-math.org
G2-M3-TE-B2-1.3.1-01.2016

125

unlabeled hundreds place value chart

Lesson 8: Count the total value of $1, $10, and $100 bills up to $1,000.

EUREKA MATH

Lesson 9

Objective: Count from $10 to $1,000 on the place value chart and the empty number line.

Suggested Lesson Structure

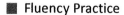

■ Fluency Practice   (15 minutes)
░ Application Problem   (8 minutes)
░ Concept Development   (30 minutes)
■ Student Debrief   (7 minutes)
   Total Time   **(60 minutes)**

Fluency Practice (15 minutes)

- Count and Change Coins to 30 Cents **2.MD.8**   (3 minutes)
- Mixed Counting with Ones, Tens, and Hundreds from 1,000 to 0 **2.NBT.2**   (5 minutes)
- Skip-Count by Twos Beginning at 394 **2.NBT.3**   (7 minutes)

Count and Change Coins to 30 Cents (3 minutes)

Materials: (T) 11 pennies, 3 dimes

> T: (Display and label a penny and a dime.) A penny has a value of 1 cent, or 1 one. A dime has a value of 10 cents, or 1 ten.
>
> T: Let's count pennies. We'll count them by ones because they have a value of 1 cent. (Lay out 1 penny at a time as students count to 10.)
>
> S: 1, 2, 3, 4, 5, 6, 7, 8, 9, 10.
>
> T: A dime has the same value as 1 ten. At the signal, say how many pennies are in a dime.
>
> S: 10 pennies are in 1 dime.
>
> T: We've counted 10 pennies; let's change them for 1 dime.
>
> T: Let's keep going, counting on from 10. (Point to the dime, and then lay out pennies as students count to 20.)
>
> S: 10, 11, 12, 13, 14, 15, 16, 17, 18, 19, 20.
>
> T: What is the value of our coins?
>
> S: 20 cents!

NOTES ON
MULTIPLE MEANS
OF REPRESENTATION:

This simple activity is students' first formal experience using coins. For this reason, it is quite guided. In the next lesson students play again, counting higher and beginning and ending with numbers other than multiples of 10.

The key is to promote gradual independence in working with coins: "I do, we do, you do."

Lesson 9:   Count from $10 to $1,000 on the place value chart and the empty
               number line.

127

T: We've completed another ten (point to the pennies). What step can we take to reduce the number of coins but keep the value of our 20 cents the same? Turn and whisper to your partner.

S: We can change our 10 pennies for another dime.

T: (Change the 10 pennies for another dime.) Thumbs up if this was your idea.

S: (Give thumbs up.)

T: Let's keep counting. Remember to count the dimes by tens and the pennies by ones.

Continue until students have reached 30 cents and changed 10 pennies for 1 dime a third time.

Mixed Counting with Ones, Tens, and Hundreds from 1,000 to 0 (5 minutes)

Materials: (T) Bundle of one hundred, one ten, and a single stick from Lesson 1

Vary numbers in this second round. Another option is to isolate a sequence that students find particularly challenging and provide them with a minute of partner practice to count up and down the sequence as fast as possible.

Skip-Count by Twos Beginning at 394 (7 minutes)

Materials: (S) Blank piece of paper

Using a blank piece of paper and a pencil, students count by twos beginning at 394. They write numbers, counting as fast and as high as they can for one minute. "Skip-count by" follows the same energizing routine for administration as Sprints. Refer to the Directions for Administration of Sprints, which are in the Module 1 Overview.

Like Sprints, after animated correction, an extra minute for independent practice, sharing with a partner, and a brief kinesthetic exercise, students repeat the counting task. The vast majority of students immediately see improvement on the second effort. Celebrate improvement in the same way as with a Sprint.

Application Problem (8 minutes)

Sarah earns $10 each week for weeding the garden. If she saves all of the money, how many weeks will it take her to save up $150?

T: Read the problem with me.

T: Work with your partner to come up with 2 different strategies to solve this problem. (Circulate and listen.)

S: I drew circles to be the tens and skip-counted up to 150. Then, I counted and it was 15 circles. → I wrote 150 equals 1 hundred 5 tens. I know 1 hundred is the same as 10 tens plus 5 tens. That's 15 tens. → I just know 15 tens is the same as 150, so she needs 15 weeks. → I wrote 150 = 100 + 50. I know 100 equals 10 tens and 50 equals 5 tens, so the answer is 10 + 5, 15.

©2015 Great Minds. eureka-math.org
G2-M3-TE-B2-1.3.1-01.2016

T: I like the way you're using unit form and expanded form to solve. Now that you've heard other strategies, talk with your partner about the one you like best and why.

T: (Allow a few minutes.) How many weeks will it take Sarah to save up $150? Give me a complete sentence.

S: It will take Sarah 15 weeks to save $150.

T: Please write that statement on your paper.

Concept Development (30 minutes)

Counting from $776 to $900 (15 minutes)

Materials: (S) Personal white board, unlabeled hundreds place value chart (Lesson 8 Template),
10 one-dollar bills, 10 ten-dollar bills, 10 hundred-dollar bills, small resealable bag per pair

Part A: Counting by One-Dollar Bills from $776 to $900 (8 minutes)

Directions:

1. Slide the place value chart inside your personal white boards.
2. Model $776 on your place value chart.
3. Model and whisper count up to $900 by ones.
4. Change 10 one-dollar bills for 1 ten-dollar bill and 10 ten-dollar bills for 1 hundred-dollar bill as you are able.
5. Each time you change 10 bills for 1 bill, let your partner handle the money.
6. If you finish before 5 minutes are up, continue counting to 1,000.

T: (Allow students time to work.) You have counted using ones. Partners, talk to your neighboring pair. When you were counting your money, when did you change 10 bills for 1 bill? Give at least two examples.

T: (Wait for pairs to share.) What unit were you just counting by?

S: Ones.

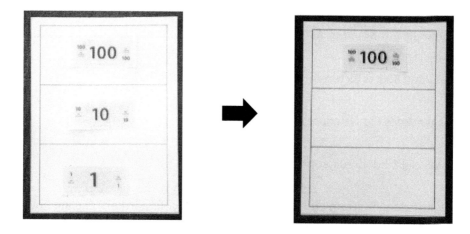

Lesson 9:   Count from $10 to $1,000 on the place value chart and the empty number line.

129

©2015 Great Minds. eureka-math.org
G2-M3-TE-B2-1.3.1-01.2016

Part B: Counting by One-Dollar, Ten-Dollar, and Hundred-Dollar Bills from $776 to $900 (7 minutes)

T: Show $776 again. This time, count up to $900 on your place value chart with one-, ten-, and hundred-dollar bills. Work with your partner to use all three units of money. If you finish early, count back down to $776.

S: (Work for two to three minutes.)

T: How did you count from 776 to 900? (Leave off the dollar signs, and record their responses as numerals, as modeled to the right, as they explain to the class.)

S: Count by ones to 780. Skip-count by tens to 800. Count one hundred to get to 900. (See ways to count to the right.)

MP.2

T: All (or both if you generate two ways) of these counts use three units, ones, tens, and hundreds in different ways.

T: Turn and talk to your partner. What are the friendly numbers in count A?

S: 780 and 800.

T: What did we count by first?

S: Ones.

T: How many ones?

S: 4 ones. (Count them if necessary.)

T: (Draw an empty number line across the board below the counts.) We started at 776 (write), and we counted up 4 ones. What number did our 4 ones get us to?

S: 780.

T: (Write 4 ones and 780 on the number line as pictured.) Next, what did we count by?

S: Tens.

T: How many tens did we skip-count?

S: 2 tens.

T: What number did 2 tens get us to?

S: 800.

T: (Write 2 tens and 800 as pictured.) Next, what did we count by?

S: Hundreds.

T: How many hundreds did we count?

S: 1 hundred.

T: What number did 1 hundred get us to?

S: 900.

T: (Write 1 hundred and 900 as pictured.) Turn and talk to your partner. Explain how the number line shows how we counted from 776 to 900. (Allow time for students to share.)

NOTES ON MULTIPLE MEANS OF ENGAGEMENT:

Push comprehension to higher levels by inviting students to analyze alternate strategies for efficiency and ease of use. These are often sweet conversations to have with students as they line up, for example: "Which did you feel was the best way to count from 776 to 900 today?"

The following are some examples of ways to count from $776 to $900 using three units:

A.   776 → 777, 778, 779, 780, 790, 800, 900

B.   776 → 876, 886, 896, 897, 898, 899, 900

C.   776 → 876, 877, 878, 879, 880, 890, 900

Lesson 9:   Count from $10 to $1,000 on the place value chart and the empty number line.

©2015 Great Minds. eureka-math.org
G2-M3-TE-B2-1.3.1-01.2016

T: Can we use this same way to count when counting bundled straws, numbers, or money?

S: Yes!

MP.2

T: A bank teller or store cashier will usually give you your change using a count like the one that we showed on our number line. Often, they count silently until they get to a friendly number. Try it.

S: (As teacher lays out bills, silently count 777, 778, 779, speak 780, silently count 790, speak 800, 900 dollars.

T: Let's try it one more time.

S: (Repeat the count.)

T: Yes. You are hired!! Bank tellers and cashiers use friendly numbers because the count is easier for the customer to follow.

Problem Set (15 minutes)

Students should do their personal best to complete the Problem Set within the allotted 15 minutes. For some classes, it may be appropriate to modify the assignment by specifying which problems they work on first. Some problems do not specify a method for solving. Students should solve these problems using the RDW approach used for Application Problems.

Directions:

1. Model the count using your money on the place value chart. Use tens and hundreds for 1–4.

2. Record your count on the empty number line Problem Set. Use our example to help you.

    1. 70 to 300.
    2. 300 to 450.
    3. 160 to 700.
    4. 700 to 870.

3. Use ones, tens, and hundreds for 5–8.

    5. 68 to 200.
    6. 200 to 425.
    7. 486 to 700.
    8. 700 to 982.

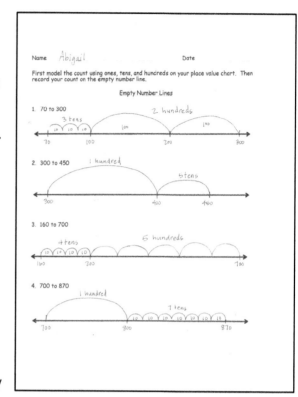

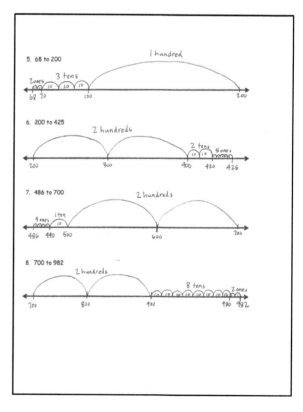

©2015 Great Minds. eureka-math.org
G2-M3-TE-B2-1.3.1-01.2016

Student Debrief (7 minutes)

Lesson Objective: Count from $10 to $1,000 on the place value chart and the empty number line.

The Student Debrief is intended to invite reflection and active processing of the total lesson experience.

Invite students to review their solutions for the Problem Set. They should check work by comparing answers with a partner before going over answers as a class. Look for misconceptions or misunderstandings that can be addressed in the Debrief. Guide students in a conversation to debrief the Problem Set and process the lesson.

T: Bring your Problem Sets with you. Go over the skip-counting you recorded on your number line. Show your partner your work, and see if you counted the same way.

T: (Circulate, watch, and listen. Resume talking after students have had time to compare and share, about one minute.)

T: Were your number lines exactly the same?

S: No!

T: There is more than one way to count, just as we saw in our lesson today. However, the number of tens and hundreds should be the same. Let's go over that.

T: Problem 1, how many hundreds did you count?

S: 2 hundreds.

T: How many tens did you count?

S: 3 tens.

T: What is the value of 2 hundreds 3 tens?

S: 230.

T: Talk to your partner. What does that 230 represent? (Circulate, watch, and listen.)

S: (After about one minute, call on students to share their thinking.)   It means that we counted 2 hundred-dollar bills and 3 ten-dollar bills when we were skip-counting from 70 to 300. → It means that on our number line first I went 3 tens and then 2 hundreds. → It means that from 70 to 300 is 230. → Yeah, it means that 300 is 230 more than 70. → It means that 70 and 230 makes 300. → Hey, it's like addition.

T: Excellent. When we put these two parts together, 70 and 230 (point to the number line, and then hold up the bills) we get $300.

T: Let's look at the next one. Did you skip-count by tens or hundreds first?

S: I counted by 1 hundred first because it's super easy to add 1 hundred to 3 hundred.

S: I skip-counted by tens first just to be different, to see what would happen.

T: How many hundreds did you count?

S: 1 hundred.

T: How many tens did you count?

S: 5 tens.

T: What is the value of 1 hundred 5 tens?

S: 150.

©2015 Great Minds. eureka-math.org
G2-M3-TE-B2-1.3.1-01.2016

T: Talk to your partner about what that 150 represents.

S: (Talk for about a minute. The students have picked up some ideas now from the first example and will be chattier now. Have them briefly share out ideas.)

T: Let's quickly go through the answers to the next two.

Repeat the process for Problems 3 and 4. Accept all number lines that make sense. It's okay if students break 5 hundreds into 5 hops on the number line. Go quickly through some answers so that the pace does not slow.

T: Today we used a tool we are very familiar with, a number line. What number lines have we used before?

S: The meter strip. → The clock. → Our number line on the classroom wall. → Our rulers are kind of like one, too.

T: How was the number line we used today different from all those other number lines? Talk to your partner.

S: It didn't have marks. → It was empty. → It didn't tell us what to count by. → We counted in different units, ones, tens, and hundreds. It made me think because I had to guess where a jump of ten was or a hundred. → Yes, it was like you just made a good guess where to draw.

T: (Listen and circulate.) I'm hearing you say that this empty number line helps you think about numbers and which jumps on the number line are bigger and which are smaller. Did it help you to model first with your money and then do it?

S: Yes!

T: Excellent. Two different ways to count! We used the number line and the place value chart. When we counted and skip-counted on the number line, as the numbers got bigger we moved from left to right (point and demonstrate silently).

T: However, when we counted on the place value chart, as the numbers got bigger we moved from right to left (demonstrate by bundling silently).

T: Turn and tell your partner how counting on the place value chart is different than on the number line.

Exit Ticket (3 minutes)

After the Student Debrief, instruct students to complete the Exit Ticket. A review of their work will help with assessing students' understanding of the concepts that were presented in today's lesson and planning more effectively for future lessons. The questions may be read aloud to the students.

Name _____   Date _____

First, model the count using ones, tens, and hundreds on your place value chart. Then, record your count on the empty number line.

Empty Number Lines

1. 70 to 300

$$\longleftrightarrow$$

2. 300 to 450

$$\longleftrightarrow$$

3. 160 to 700

$$\longleftrightarrow$$

4. 700 to 870

$$\longleftrightarrow$$

Lesson 9: Count from $10 to $1,000 on the place value chart and the empty number line.

EUREKA
MATH™

5. 68 to 200

6. 200 to 425

7. 486 to 700

8. 700 to 982

Lesson 9:   Count from $10 to $1,000 on the place value chart and the empty
number line.

135

©2015 Great Minds. eureka-math.org
G2-M3-TE-B2-1.3.1-01.2016

Name _____   Date _____

1. Jeremy counted from $280 to $435. Use the number line to show a way that Jeremy could have used ones, tens, and hundreds to count.

←――――――――――――――――――――――――――――――→

2. Use the number line to show another way that Jeremy could have counted from $280 to $435.

←――――――――――――――――――――――――――――――→

3. Use the number line to show how many hundreds, tens, and ones you use when you count from $776 to $900.

←――――――――――――――――――――――――――――――→

     To count from $776 to $900, I used _____ hundreds _____ tens _____ ones.

Lesson 9:   Count from $10 to $1,000 on the place value chart and the empty number line.

EUREKA MATH™

Name _____ Date _____

1. Write the total amount of money shown in each group.

a.

$100	$100
$100	$100
$100	$100
$100	$100
$100	$100

_____

b.

$10	$10
$10	$10
$10	$10
$10	$10
$10	$10

_____

c.

$1	$1
$1	$1
$1	$1
$1	$1
$1	$1

_____

d.

$10	$100
$10	$100
$10	$100
$100	$1
$100	$1

_____

2. Show one way to count from $82 to $512.

EUREKA MATH

Lesson 9:   Count from $10 to $1,000 on the place value chart and the empty number line.

137

©2015 Great Minds. eureka-math.org
G2-M3-TE-B2-1.3.1-01.2016

3. Use each number line to show a different way to count from $580 to $994.

4. Draw and solve.
   Julia wants a bike that costs $75. She needs to save $25 more to have enough money to buy it. How much money does Julia already have?

   Julia already has $_____.

Count from $10 to $1,000 on the place value chart and the empty number line.

EUREKA
MATH™

Lesson 10

Objective: Explore $1,000. How many $10 bills can we change for a thousand dollar bill?

Suggested Lesson Structure

- Fluency Practice   (14 minutes)
- Application Problem   (31 minutes)
- Student Debrief   (15 minutes)
  Total Time   **(60 minutes)**

Fluency Practice (14 minutes)

- Count and Change Coins from 85 to 132 Cents **2.NBT.8**   (3 minutes)
- Sprint: More Expanded Form **2.NBT.3**   (8 minutes)
- Skip-Count by Tens: Up and Down Between 0 and 1,000 **2.NBT.2**   (3 minutes)

Count and Change Coins from 85 to 132 Cents (3 minutes)

Materials: (T) 16 pennies and 13 dimes

  T: (Display and label a penny and a dime.) At the signal, say the answer. A penny is like 1 one, 1 ten, or 1 hundred?
  S: 1 one!
  T: A dime is like 1 one, 1 ten, or 1 hundred?
  S: 1 ten!
  T: Let's count. (Quickly lay out 85 cents using 8 dimes and 5 pennies.)
  S: 10, 20, 30, 40, 50, 60, 70, 80, 81, 82, 83, 84, 85.
  T: (Lay out another dime.) Whisper the new value of our money to your partner. (Take note of students who have difficulty with this.)
  S: 95 cents.
  T: Let's count on. (Lay out pennies as students count to 105.)
  S: 96, 97, 98, 99, 100, 101, 102, 103, 104, 105.

> **NOTES ON MULTIPLE MEANS OF ENGAGEMENT:**
>
> Coin names are important and take time for English language learners to learn. It is wise to have a classroom economy (search online under *classroom economies for children*) using coins so that they are used again and again. Repetition is crucial for language acquisition. There are many suggestions online that meet the needs of diverse classroom cultures.

Lesson 10:   Explore $1,000. How many $10 bills can we change for a thousand dollar bill?

139

©2015 Great Minds. eureka-math.org
G2-M3-TE-B2-1.3.1-01.2016

T: The new value of our money is…?

S: 105 cents!

T: Whisper to your partner how we can reduce the number of coins but keep the value the same.

S: Change 10 pennies for a dime. (Take note of students who are uncertain, possibly because 105 is not a multiple of 10.)

T: (Continue, mixing counting by ones and tens to 125. Vary the practice in response to noticing where students have difficulty in the first counts. Remember to count from 125 to 132 using pennies.)

Sprint: More Expanded Form (8 minutes)

Materials: (S) More Expanded Form Sprint

Skip-Count by Tens: Up and Down Between 0 and 1,000 (3 minutes)

T: Let's play Happy Counting skip-counting by tens!

T: Watch my fingers to know whether to count up or down. A closed hand means stop. (Show signals while explaining.)

T: Let's count up by tens, starting at 560. Ready? (Rhythmically point up until a change is desired. Show a closed hand, and then point down. Continue, mixing it up.)

S: 560, 570, 580, 590, 600, 610, 620 (stop). 610, 600 (stop). 610, 620, 630, 640, 650, 660, 670, 680, 690 (stop). 680, 670, 660 (stop). 670, 680, 690, 700, 710, 720, 730 (stop). 720, 710, 700.

Application Problem (31 minutes)

Materials: (S) Problem Set (if unable to project during the Debrief, perhaps have the students do their work on posters rather than 8 ½" x 11" paper)

T: Read the following story:

Jerry is a second grader. He was playing in the attic and found an old, dusty trunk. When he opened it, he found things that belonged to his grandfather. There was a cool collection of old coins and bills in an album. One bill was worth $1,000. Wow! Jerry lay down and started daydreaming. He thought about how good it would feel to give as many people as he could a ten-dollar bill. He thought about how he had felt on his birthday last year when he got a card from his uncle with a ten-dollar bill inside.

NOTES ON
MULTIPLE MEANS
OF ACTION AND
EXPRESSION:

To support understanding, the story can be acted out or illustrated. Never underestimate the increased comprehension offered by the simplest of illustrations. A hand-drawn thousand-dollar bill would be a useful prop for acting out the story. Also, ask students questions such as, "What would you do if you found $1,000? Tell your partner."

But even more, he thought about how lucky he felt one snowy, cold day walking to school when he found a ten-dollar bill in the snow. Maybe he could quietly hide the ten-dollar bills so that lots of people could feel as lucky as he did on that cold day! He thought to himself, "I wonder how many ten-dollar bills are equal to a thousand-dollar bill? I wonder how many people I could bring a lucky day to?"

Lesson 10: Explore $1,000. How many $10 bills can we change for a thousand dollar bill?

T: Summarize the story to your partner from the beginning to the end the best you can.

T: (After students talk for about a minute, it is clear whether they can reconstruct the story. Invite them to listen once again to fill in missing details if necessary.)

T: You will work in pairs to answer Jerry's question. What is his question?

S: To know how many people he can give a ten-dollar bill to. → To find out how many ten-dollar bills are the same as a thousand-dollar bill.

T: At the end of 20 minutes, you will put your work on your table, and we will do a gallery walk so that you will have a chance to see everyone's work.

T: (Pass out the Problem Set.) Let's go over the directions.

T: Answer Jerry's question: "I wonder how many ten-dollar bills are equal to a thousand-dollar bill?" Use the RDW strategy, and explain your solution using words, pictures, or numbers.

T: Work with your partner to solve the problem. Use a full sheet of paper. Remember to write your answer in a statement.

Note: As the students work, ask them to think about the tools and strategies they have learned and used thus far in the year. Much of MP.1's "Make sense of problems and persevere in solving them" and MP.5's "Use appropriate tools strategically" involves encouraging students to move through indecision and not knowing to making choices independently. Encourage them to try what comes: "Go for it," or, "See if it works." Often, students start to strategize when they realize a choice is ineffective. This is a day to let that happen. Make an effort to sit back and watch your class objectively. Make notes on who is struggling. Notice what their partner does in response. Notice how they re-engage. If a student loses focus, consider some simple focus questions such as, "What is the problem asking you?" Or, "Is your pencil sharp enough?" Redirection can be quick and subtle but effective.

Do give students time signals: "You have 10 minutes," or, "You have 5 minutes." For students who succeed quickly, post a challenge problem, such as Jerry's grandfather took the thousand-dollar bill to the bank and changed it for some ten-dollar and hundred-dollar bills. If he gave Jerry and his sister each one hundred dollars, how much money will he have left?

Student Debrief (15 minutes)

Lesson Objective: Explore $1,000. How many $10 bills can we change for a thousand dollar bill?

Materials: (S) Completed Problem Set

T: Let's set up our gallery in order to allow everyone to see each other's work. Pair 1, place your work here. Pair 2, place your work here. (Arrange the work in a circular configuration.)

T: When I give the signal, advance to the next table to your right. Make note of the different strategies your friends have used to solve Jerry's problem.

Keep this process moving. It will take about 5–7 minutes for 10 pairs to move through the "gallery."

T: Bring your work to the carpet for our Debrief.

T: Do we all agree how many ten-dollar bills Jerry will be able to share? Tell me at the signal.

S: 100 ten-dollar bills!

EUREKA
MATH™

Lesson 10: Explore $1,000. How many $10 bills can we change for a thousand dollar bill?

141

©2015 Great Minds. eureka-math.org
G2-M3-TE-B2-1.3.1-01.2016

T: Most groups were able to come up with that answer, but did everyone's work look the same?

S: No!

T: You used different strategies. Let's look more closely at some different ways of solving the problem.

Analysis of One Piece of Student Work

T: I would like to start out looking at Brandon, Pedro, and Wanda's work. (Post the student work.)

T: When you look at their work, talk to your partner about what you see. (Circulate and listen.)

S: I see number bonds. → I see that a number bond has hundreds sticking out. → I see the other number bond is different. It has 10 tens instead of hundreds.

T: Does anyone have a quick compliment for this team's math?

S: I would like to compliment that you made it easy for us to see the number of tens.

S: I like that you showed how you counted up the total number of tens. That was easy to understand.

T: Good compliments. Does anyone have a suggestion or a question?

S: A question I have is why did you draw two number bonds?

S: Well, we didn't know the answer, but we knew that we could skip-count by hundreds up to one thousand. So, we just started drawing and counting. Then, Wanda said that we could make another one and that we could write the tens instead.

T: So, they got going and got another idea. Excellent. Let's remember that! Don't get too stuck.
Try something. Suggestions? Questions?

S: I think if the two bonds were the same size, it would be even easier to see that 100 is 10 tens.

S: I have an idea. Maybe when you are counting up the tens, you could write 10 tens, 20 tens, 30 tens right inside the parts.

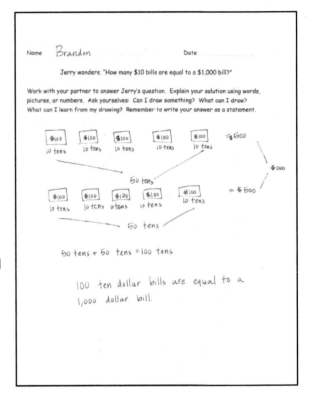

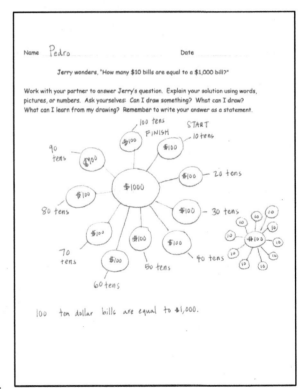

Lesson 10: Explore $1,000. How many $10 bills can we change for a thousand dollar bill?

Comparison of Two Pieces of Student Work

T: Let's look at the work of Sammy, Olga, and Marisela. Talk to your team again. What do you see?

S: They made a number line. → I see they counted up to 1,000 by skip-counting by 1 hundred. → I see that each hop has 10 tens written inside it. → There are 10 hops in all. → They have counted by ten under the number line right here. Maybe that's where they were figuring out how many tens in all.

T: Let's compare the number bond team and the number line team's solutions. Talk to your partner. What is different about the way they represented the problem and what is the same?

S: Both of them got the right answer. → I like the number line better. → She didn't say to talk about what we liked, just what was the same and different. → They both count by tens and 10 tens. → And they both skip-counted by hundreds.

MP.2

T: So, two different tools, a number bond and a number line. Now look at Freddy, Vincent, and Eva's work. What tool did they use?

S: Ten-frames!

T: Compare the way they used the ten-frames with the way the other team used the number bonds.

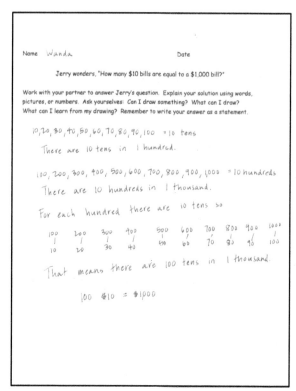

Continue the math talk, asking students to compare the representations.

For example:

- Freddy, Vincent, and Eva used the ten-frame but only drew 5 of them. How did they get the right answer?
- Where on the number bond and number line work do you see the 5 ten-frames?
- What are the advantages of using the number line?
- What are the advantages of using the number bond?

Be sure to get students to realize that drawing all the bills takes a long time but that the idea of drawing half is wise.

Exit Ticket (3 minutes)

After the Student Debrief, instruct students to complete the Exit Ticket. A review of their work will help with assessing students' understanding of the concepts that were presented in today's lesson and planning more effectively for future lessons. The questions may be read aloud to the students.

Lesson 10:   Explore $1,000. How many $10 bills can we change for a thousand dollar bill?

143

©2015 Great Minds. eureka-math.org
G2-M3-TE-B2-1.3.1-01.2016

A

Number Correct: _____

Expanded Form

1.	100 + 20 + 3 =	
2.	100 + 20 + 4 =	
3.	100 + 20 + 5 =	
4.	100 + 20 + 8 =	
5.	100 + 30 + 8 =	
6.	100 + 40 + 8 =	
7.	100 + 70 + 8 =	
8.	500 + 10 + 9 =	
9.	500 + 10 + 8 =	
10.	500 + 10 + 7 =	
11.	500 + 10 + 3 =	
12.	700 + 30 =	
13.	700 + 3 =	
14.	30 + 3 =	
15.	700 + 33 =	
16.	900 + 40 =	
17.	900 + 4 =	
18.	40 + 4 =	
19.	900 + 44 =	
20.	800 + 70 =	
21.	800 + 7 =	
22.	70 + 7 =	

23.	800 + 77 =	
24.	300 + 90 + 2 =	
25.	400 + 80 =	
26.	600 + 7 =	
27.	200 + 60 + 4 =	
28.	100 + 9 =	
29.	500 + 80 =	
30.	80 + 500 =	
31.	2 + 50 + 400 =	
32.	2 + 400 + 50 =	
33.	3 + 70 + 800 =	
34.	40 + 9 + 800 =	
35.	700 + 9 + 20 =	
36.	5 + 300 =	
37.	400 + 90 + 10 =	
38.	500 + 80 + 20 =	
39.	900 + 60 + 40 =	
40.	400 + 80 + 2 =	
41.	300 + 60 + 5 =	
42.	200 + 27 + 5 =	
43.	8 + 700 + 59 =	
44.	47 + 500 + 8 =	

Lesson 10: Explore $1,000. How many $10 bills can we change for a thousand dollar bill?

EUREKA
MATH™

©2015 Great Minds. eureka-math.org
G2-M3-TE-B2-1.3.1-01.2016

B

Number Correct: _____

Improvement: _____

Expanded Form

1.	100 + 30 + 4 =	
2.	100 + 30 + 5 =	
3.	100 + 30 + 6 =	
4.	100 + 30 + 9 =	
5.	100 + 40 + 9 =	
6.	100 + 50 + 9 =	
7.	100 + 80 + 9 =	
8.	400 + 10 + 8 =	
9.	400 + 10 + 7 =	
10.	400 + 10 + 6 =	
11.	400 + 10 + 2 =	
12.	700 + 80 =	
13.	700 + 8 =	
14.	80 + 8 =	
15.	700 + 88 =	
16.	900 + 20 =	
17.	900 + 2 =	
18.	20 + 2 =	
19.	900 + 22 =	
20.	700 + 60 =	
21.	700 + 6 =	
22.	60 + 6 =	

23.	700 + 66 =	
24.	200 + 90 + 4 =	
25.	500 + 70 =	
26.	800 + 6 =	
27.	400 + 70 + 4 =	
28.	700 + 9 =	
29.	800 + 50 =	
30.	50 + 800 =	
31.	2 + 80 + 400 =	
32.	2 + 400 + 80 =	
33.	3 + 70 + 500 =	
34.	60 + 3 + 800 =	
35.	900 + 7 + 20 =	
36.	4 + 300 =	
37.	500 + 90 + 10 =	
38.	600 + 80 + 20 =	
39.	900 + 60 + 40 =	
40.	600 + 8 + 2 =	
41.	800 + 6 + 5 =	
42.	800 + 27 + 5 =	
43.	8 + 100 + 49 =	
44.	37 + 600 + 8 =	

EUREKA MATH™

Lesson 10:   Explore $1,000. How many $10 bills can we change for a thousand dollar bill?

145

Name _____   Date _____

Jerry wonders, "How many $10 bills are equal to a $1,000 bill?"

Work with your partner to answer Jerry's question. Explain your solution using words, pictures, or numbers. Ask yourselves: Can I draw something? What can I draw? What can I learn from my drawing? Remember to write your answer as a statement.

Lesson 10:   Explore $1,000. How many $10 bills can we change for a thousand dollar bill?

Name _____   Date _____

Jerry wonders, "How many $10 bills are equal to a $1,000 bill?"

Think about the different strategies your classmates used to answer Jerry's question. Answer the problem again using a strategy you liked that is <u>different</u> from yours. Use words, pictures, or numbers to explain why that strategy also works.

Lesson 10:   Explore $1,000. How many $10 bills can we change for a thousand
                dollar bill?

©2015 Great Minds. eureka-math.org
G2-M3-TE-B2-1.3.1-01.2016

147

Name _____ Date _____

Jerry wonders, "How many $10 bills are equal to a $1,000 bill?"

Think about the strategies your friends used to answer Jerry's question. Answer the problem again using a different strategy than the one you used with your partner and for the Exit Ticket. Explain your solution using words, pictures, or numbers. Remember to write your answer as a statement.

Lesson 10:   Explore $1,000. How many $10 bills can we change for a thousand dollar bill?

Name _____   Date _____

1. Dora has saved $314.

   a. Write the amount Dora has saved in three different ways by filling in the blanks.

      word form _____

      expanded form _____

      __ hundreds __ tens __ ones

   b. Dora's goal is to save $400. How many tens are in $400? Explain your answer using words, pictures, or numbers.

c. Dora reaches her goal of $400 in savings. She decides to set a new goal of $900. How many more
    $100 bills will she need to reach $900 in savings? Explain your answer using words, pictures, or
    numbers.

d. Dora made her new goal! She saved both ten-dollar bills and hundred-dollar bills to go from $400 to
    $900. Show how Dora could skip-count using tens **and** hundreds from 400 to 900. Explain your
    answer using words, pictures, or numbers.

EUREKA
MATH™

©2015 Great Minds. eureka-math.org
G2-M3-TE-B2-1.3.1-01.2016

Mid-Module Assessment Task Standards Addressed	Topics A–D

Understand place value.

2.NBT.1 Understand that the three digits of a three-digit number represent amounts of hundreds, tens, and ones; e.g., 706 equals 7 hundreds, 0 tens, and 6 ones. Understand the following as special cases:

　　a.　100 can be thought of as a bundle of ten tens—called a "hundred."

　　b.　The numbers 100, 200, 300, 400, 500, 600, 700, 800, 900 refer to one, two, three, four, five, six, seven, eight, or nine hundreds (and 0 tens and ones).

2.NBT.2 Count within 1000: skip-count by 5s, 10s and 100s.

2.NBT.3 Read and write numbers to 1000 using base-ten numerals, number names, and expanded form.

Evaluating Student Learning Outcomes

A Progression Toward Mastery is provided to describe steps that illuminate the gradually increasing understandings that students develop on their way to proficiency. In this chart, this progress is presented from left (Step 1) to right (Step 4). The learning goal for students is to achieve Step 4 mastery. These steps are meant to help teachers and students identify and celebrate what the students CAN do now and what they need to work on next.

©2015 Great Minds. eureka-math.org
G2-M3-TE-B2-1.3.1-01.2016

A Progression Toward Mastery

Assessment Task Item and Standards Assessed	STEP 1 Little evidence of reasoning without a correct answer. (1 Point)	STEP 2 Evidence of some reasoning without a correct answer. (2 Points)	STEP 3 Evidence of some reasoning with a correct answer or evidence of solid reasoning with an incorrect answer. (3 Points)	STEP 4 Evidence of solid reasoning with a correct answer. (4 Points)
1(a) 2.NBT.1 2.NBT.3	Student is not able to accurately represent hundreds, tens, and ones.	Student shows evidence of beginning to represent 314, but the solution is incorrect for two of the three answers.	Student understands how to represent 314 correctly for two of the three answers.	Student correctly represents three ways of writing 314: • Three hundred fourteen • 300 + 10 + 4 = 314 • 3 hundreds 1 ten 4 ones
1(b) 2.NBT.1a	Student is not able to decide on a strategy or is not able to count accurately by tens.	Student shows evidence of beginning to use a counting strategy but is unable to get the right answer.	Student has the correct answer of 40 but is unable to explain accurately using pictures, numbers, or words to clearly demonstrate reasoning. OR Student is able to show skip-counting or a bundling strategy but has an incorrect answer.	Student uses an accurate counting strategy, with the correct answer of 40, and gives a clear explanation using pictures, numbers, and/or words.
1(c) 2.NBT.1b	Student is not able to decide on a strategy or is not able to count accurately by hundreds.	Student shows evidence of beginning to use a counting strategy but has an incorrect answer.	Student has the correct answer but is unable to show sound counting or reasoning. OR Student is able to reason counting by hundreds but with an incorrect answer.	Student counts correctly by hundreds with a correct answer of 5 hundred-dollar bills, showing reasoning using pictures, numbers, and/or words.

EUREKA MATH™

©2015 Great Minds. eureka-math.org
G2-M3-TE-B2-1.3.1-01.2016

A Progression Toward Mastery

1(d) 2.NBT.1 2.NBT.2	Student is not able to decide on a strategy or is not able to count accurately by tens and hundreds.	Student shows evidence of beginning to count by tens and/or by hundreds but is unable to use both to reach a correct answer.	Student has a correct answer but does not clearly demonstrate an answer that uses both tens and hundreds. OR Student has an incorrect answer but demonstrates clearly.	Student uses tens and hundreds to count correctly from $400 to $900, using skip-counting or bundling in pictures, numbers, and/or words.

Name   Freddy   Date

1. Dora has saved $314.

   a. Write the amount Dora has saved in three different ways by filling in the blanks.

   word form   three hundred fourteen

   expanded form   300 + 10 + 4 = 314

   3 hundreds   1 tens   4 ones

   b. Dora's goal is to save $400. How many tens are in $400? Explain your answer using words, pictures or numbers.

   40 tens are inside 400. You can see in the picture how I counted. Also, unit form. 40 tens 0 ones is 400.

EUREKA MATH

c.   Dora reaches her goal of $400 in savings. She decides to set a new goal of $900. How many more
$100 bills will she need to reach $900 in savings? Explain your answer using words, pictures, or
numbers.

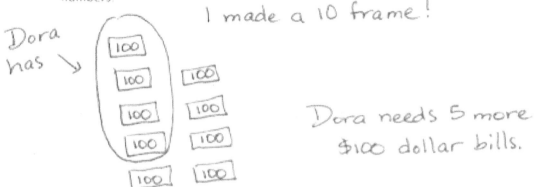

I made a 10 frame!

Dora
has →

Dora needs 5 more
$100 dollar bills.

d.   Dora made her new goal! She saved both ten dollar bills and hundred dollar bills to go from $400 to
$900. Show how Dora could skip-count using tens and hundreds from 400 to 900. Explain your
answer using words, pictures, or numbers.

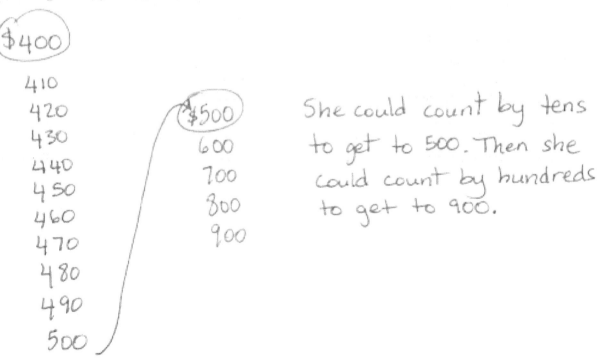

$400

410
420
430
440
450
460
470
480
490
500

$500
600
700
800
900

She could count by tens
to get to 500. Then she
could count by hundreds
to get to 900.

©2015 Great Minds. eureka-math.org
G2-M3-TE-B2-1.3.1-01.2016

2 GRADE

Mathematics Curriculum

Topic E

Modeling Numbers Within 1,000 with Place Value Disks

2.NBT.A

Focus Standard:	2.NBT.A	Understand place value.
Instructional Days:	5	
Coherence -Links from:	G1–M6	Place Value, Comparison, Addition and Subtraction to 100
-Links to:	G2–M4	Addition and Subtraction Within 200 with Word Problems to 100
	G2–M7	Problem Solving with Length, Money, and Data

In Topic E, students transition to the more abstract place value disks that are used through Grade 5 for modeling very large and very small numbers. The foundation has been carefully laid for this moment since Kindergarten, when students first learned how much a number less than 10 needs to make ten. Students repeat the counting lessons of the bundles and money but with place value disks (**2.NBT.2**).

The three representations—bundles, money, and disks—each play an important role in students' deep internalization of the meaning of each unit on the place value chart (**2.NBT.1**). Like bills, disks are "traded," "renamed," or "changed for" a unit of greater value (**2.NBT.2**).

Finally, students evaluate numbers in unit form with more than 9 ones or tens, for example, 3 hundreds 4 tens 15 ones and 2 hundreds 15 tens 5 ones. Topic E also culminates with a problem-solving exploration in which students use counting strategies to solve problems involving pencils that come in boxes of 10 (**2.NBT.2**).

A Teaching Sequence Toward Mastery of Modeling Numbers Within 1,000 with Place Value Disks

Objective 1: Count the total value of ones, tens, and hundreds with place value disks.
(Lesson 11)

Objective 2: Change 10 ones for 1 ten, 10 tens for 1 hundred, and 10 hundreds for 1 thousand.
(Lesson 12)

Objective 3: Read and write numbers within 1,000 after modeling with place value disks.
(Lesson 13)

Objective 4: Model numbers with more than 9 ones or 9 tens; write in expanded, unit, standard, and word forms.
(Lesson 14)

Objective 5: Explore a situation with more than 9 groups of ten.
(Lesson 15)

Lesson 11

Objective: Count the total value of ones, tens, and hundreds with place value disks.

Suggested Lesson Structure

■ Fluency Practice   (12 minutes)
■ Application Problem   (9 minutes)
■ Concept Development   (25 minutes)
■ Student Debrief   (14 minutes)

   Total Time   **(60 minutes)**

Fluency Practice (12 minutes)

- Rekenrek Counting: Numbers in Unit Form Between 11 and 100 **2.NBT.1**   (4 minutes)
- Sprint: Addition and Subtraction to 10 **2.OA.2**   (8 minutes)

Rekenrek Counting: Numbers in Unit Form Between 11 and 100 (4 minutes)

Materials: (T) Rekenrek

   T: (Show 11.) What number is showing?
   S: 11.
   T: The unit form way?
   S: 1 ten 1 one.
   T: Good. Keep counting the unit form way. (Move beads to count by ones to 15.)
   S: 1 ten 2 ones, 1 ten 3 ones, 1 ten 4 ones, 1 ten 5 ones.
   T: This time say each number two ways. First, the unit form way, then just as ones. Let's do one together so you know what I mean. (Switch to counting by tens, and show 25.)
   T: Me first. 2 tens 5 ones is 25 ones. Your turn.
   S: 2 tens 5 ones is 25 ones.
   T: Good. Say the numbers that I show both ways. (Continue to count by tens to 55.)
   S: 3 tens 5 ones is 35 ones, 4 tens 5 ones is 45 ones, 5 tens 5 ones is 55 ones.
   T: This time say the ones first, and then the unit form. (Switch to counting by ones to 61.)
   S: 56 ones is 5 tens 6 ones, 57 ones is 5 tens 7 ones, 58 ones is 5 tens 8 ones, 59 ones is 5 tens 9 ones, 60 ones is 6 tens, 61 ones is 6 tens 1 one.

Continue with the following possible sequence: Count down by tens from 97 to 37, and count down by ones from 37 to 25.

Lesson 11:   Count the total value of ones, tens, and hundreds with place value disks.

©2015 Great Minds. eureka-math.org
G2-M3-TE-B2-1.3.1-01.2016

Sprint: Addition and Subtraction to 10 (8 minutes)

Materials: (S) Addition and Subtraction to 10 Sprint

Application Problem (9 minutes)

Samantha is helping the teacher organize the pencils in her classroom. She finds 41 yellow pencils and 29 blue pencils. She throws away 12 that are too short. How many pencils are left in all?

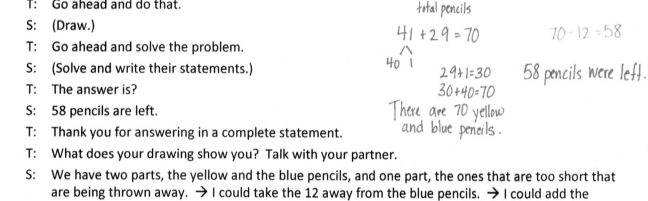

T: When you read this story, what do you see?

S: Pencils. → Yellow and blue pencils. → 12 pencils that are too short.

T: Can you draw something to represent the pencils?

S: We can draw the pencils. → We can draw bundles. → We can draw boxes of 10 pencils.

T: I'm only giving you two minutes to draw, so would it be wiser to draw bundles, boxes, or all of the pencils?

S: Bundles or boxes.

T: Go ahead and do that.

S: (Draw.)

T: Go ahead and solve the problem.

S: (Solve and write their statements.)

T: The answer is?

S: 58 pencils are left.

T: Thank you for answering in a complete statement.

T: What does your drawing show you? Talk with your partner.

S: We have two parts, the yellow and the blue pencils, and one part, the ones that are too short that are being thrown away. → I could take the 12 away from the blue pencils. → I could add the yellow and blue pencils and take away the short ones from the total. → I could take the short ones away from the yellow pencils and then add the blue. → Yeah, that's true because even though it was maybe a mix of blue and yellow ones that were too short, it still will tell the right total in the end.

T: Let's look at two different work samples that solved the problem in different ways.

Concept Development (25 minutes)

Materials: (T) Dienes blocks (9 hundreds, 9 tens, 9 ones), unlabeled hundreds place value chart (Lesson 8 Template), place value disks (9 hundreds, 9 tens, 9 ones) (S) Dienes blocks (2 hundreds, 9 tens, 9 ones) unlabeled hundreds place value chart (Lesson 8 Template), place value disks (6 hundreds, 9 tens, 9 ones) place value disks (Template)

T: Slide the place value chart inside your personal white boards.

T: With your blocks, show me this number. (Silently write 13 on the board.)

S: (Show.)

T: Whisper the number first in unit form and then in standard form.

S: 1 ten 3 ones, thirteen.

T: (Point to place value disks.) Show me the same number with your **place value disks,** and whisper the unit form and standard form as you work.

S: (Show and whisper.)

T: With your blocks, show the number to me. (Silently write 103 on the board.)

S: (Show.)

T: Whisper the number first in unit form and then in standard form.

S: 1 hundred 3 ones, one hundred three.

T: Show me the same number with your place value disks, and whisper as you work.

S: (Show.)

Continue alternating between blocks and disks, possibly with the following sequence: 129, 130, 230, 203, 199, 200. (For now, please resist using the words *more* or *less*.)

T: Talk with your partner about the difference between modeling your numbers with blocks and modeling your numbers with place value disks.

S: The blocks were yellow, and the place value disks were different colors. → The blocks were bigger and smaller, and the place value disks were all the same size. → The place value disks have the name on them. The blocks don't. You just count.

T: Up to this point, we have been using bundles (hold up 1 hundred) and bills (hold up 1 hundred-dollar bill). Talk to your partner, and compare the blocks and place value disks to the bundles and bills. How are they the same? How are they different?

S: The bills have the name on them like the place value disks. → With the bundles you can count the number of straws like the blocks. → The bundles and blocks both are bigger when you have a bigger number.
→ The bills and place value disks stay the same size.
→ They all represent hundreds, tens, and ones.
→ The bills and the straws we see at home, but these blocks and place value disks are just in math class. I've never seen them anywhere else.

T: Okay, as I am circulating and listening, I hear some very thoughtful insights.

NOTES ON
MULTIPLE MEANS
OF ACTION AND
EXPRESSION:

In order to promote deeper discussion, vary the grouping in the classroom, such that students who speak the same native languages are grouped together. This works particularly well when the discussion is open-ended and invites students to reflect on their own learning through the context of a hypothetical situation. Importantly, it does not require translation. Key vocabulary is familiar and supported with visuals, and it is secondary in purpose to the thinking that students are asked to produce.

©2015 Great Minds. eureka-math.org
G2-M3-TE-B2-1.3.1-01.2016

T: Here is a question to discuss with your partner. Imagine you are a teacher. How would you use these tools to teach different things to your class? (Write or post each word with a small pictorial for each to support language use.)

- Bundles
- Blocks
- Bills
- Place value disks

Problem Set (10 minutes)

Students should do their personal best to complete the Problem Set within the allotted 10 minutes. For some classes, it may be appropriate to modify the assignment by specifying which problems they work on first. Some problems do not specify a method for solving. Students should solve these problems using the RDW approach used for Application Problems.

Problem 1(a–e): Model the numbers on your place value chart using the fewest number of blocks or disks possible.

1. Partner A, use base ten blocks.

2. Partner B, use place value disks.

3. Whisper each number in unit and standard form.

Problem 2(a–j): Model the numbers on your place value chart using the fewest number of place value disks possible.

1. Partners A and B alternate using place value disks.

2. Whisper each number in unit and standard form.

Student Debrief (14 minutes)

Lesson Objective: Count the total value of ones, tens, and hundreds with place value disks.

The Student Debrief is intended to invite reflection and active processing of the total lesson experience.

Invite students to review their solutions for the Problem Set. They should check work by comparing answers with a partner before going over answers as a class. Look for misconceptions or misunderstandings that can be addressed in the Debrief. Guide students in a conversation to debrief the Problem Set and process the lesson.

NOTES ON MULTIPLE MEANS OF ENGAGEMENT:

This Problem Set lends itself well to pairing up accelerated and struggling students. Encourage students with varying skill levels or levels of English language competence to teach and assist one another when building the models. This cultivates a classroom community that thrives on mutual support and cooperation. As students work together to solve problems, monitor their progress to ensure that everyone is engaged and participating.

Name Ashlyn   Date

1. Model the numbers on your place value chart using the fewest number of blocks or disks possible.

    Partner A, use base ten blocks.
    Partner B, use place value disks.
    Compare the way your numbers look.
    Whisper the numbers in standard form and unit form.

    a. 12
    b. 124
    c. 104
    d. 299
    e. 200

2. Take turns using the place value disks to model the following numbers using the fewest disks possible. Whisper the numbers in standard form and unit form.

    a. 25   f. 36
    b. 250   g. 360
    c. 520   h. 630
    d. 502   i. 603
    e. 205   j. 306

Lesson 11:   Count the total value of ones, tens, and hundreds with place value disks.

161

©2015 Great Minds. eureka-math.org
G2-M3-TE-B2-1.3.1-01.2016

T: Come to the carpet with your partner and your Problem Set. Whisper skip-count down by tens from 300 as you transition to the carpet.

T: Let's begin with Problem 2(a) and (b). Discuss with your partner how the numbers changed using this sentence frame (posted or written).

I changed _____ to _____.

I changed _____ to _____.

The value of my number changed from _____ to _____.

S: (Might catch on quickly.) I changed 2 tens to 2 hundreds. I changed 5 ones to 5 tens. The value of my number changed from 25 to 250.

T: (If not, ask a student to model.) Let's have Alejandro use his words for us.

S: I changed 2 tens to 2 hundreds. I changed 5 ones to 5 tens. That changed the value of my number from 25 to 250.

T: Just as Alejandro demonstrated, tell your partner how the numbers changed from Problem 2(b) to 2(c).

S: I changed 2 hundreds to 5 hundreds. I changed 5 tens to 2 tens. The value of my number changed from 250 to 520.

T: You improved! Keep going through the Problem Set's numbers using words to tell about the changes. (Continue for about four minutes while circulating and supporting.)

T: Today we used a new tool, **place value disks**. Did you enjoy using them?

S: Yes!

T: We will keep our bundles of straws and our base ten blocks here in the math materials center. They will always help us remember the value of our units. I will hold up a unit, and you show me the correct place value disk.

T: (Silently hold up a flat. Students hold up a hundred disk. Hold up a bundle of 10 straws. Students hold up a ten disk. Hold up a one-dollar bill, etc.)

T: Quietly go back to your seats to complete your Exit Ticket.

Exit Ticket (3 minutes)

After the Student Debrief, instruct students to complete the Exit Ticket. A review of their work will help with assessing students' understanding of the concepts that were presented in today's lesson and planning more effectively for future lessons. The questions may be read aloud to the students.

Note: Students need the place value disks template to complete homework. Students may cut it apart and store the disks in a small resealable bag for use at home.

Lesson 11:   Count the total value of ones, tens, and hundreds with place value disks.

©2015 Great Minds. eureka-math.org
G2-M3-TE-B2-1.3.1-01.2016

A

Number Correct: _____

Addition and Subtraction to 10

1.	2 + 1 =	
2.	1 + 2 =	
3.	3 – 1 =	
4.	3 – 2 =	
5.	4 + 1 =	
6.	1 + 4 =	
7.	5 – 1 =	
8.	5 – 4 =	
9.	8 + 1 =	
10.	1 + 8 =	
11.	9 – 1 =	
12.	9 – 8 =	
13.	3 + 2 =	
14.	2 + 3 =	
15.	5 – 2 =	
16.	5 – 3 =	
17.	5 + 2 =	
18.	2 + 5 =	
19.	7 – 2 =	
20.	7 – 5 =	
21.	6 + 2 =	
22.	2 + 6 =	

23.	8 – 2 =	
24.	8 – 6 =	
25.	8 + 2 =	
26.	2 + 8 =	
27.	10 – 2 =	
28.	10 – 8 =	
29.	4 + 3 =	
30.	3 + 4 =	
31.	7 – 3 =	
32.	7 – 4 =	
33.	5 + 3 =	
34.	3 + 5 =	
35.	8 – 3 =	
36.	8 – 5 =	
37.	6 + 3 =	
38.	3 + 6 =	
39.	9 – 3 =	
40.	9 – 6 =	
41.	5 + 4 =	
42.	4 + 5 =	
43.	9 – 5 =	
44.	9 – 4 =	

EUREKA MATH™

Lesson 11:   Count the total value of ones, tens, and hundreds with place value disks.

163

©2015 Great Minds. eureka-math.org
G2-M3-TE-B2-1.3.1-01.2016

B

Addition and Subtraction to 10

Number Correct: _____

Improvement: _____

1.	3 + 1 =		23.	7 – 2 =	
2.	1 + 3 =		24.	7 – 5 =	
3.	4 – 1 =		25.	8 + 2 =	
4.	4 – 3 =		26.	2 + 8 =	
5.	5 + 1 =		27.	10 – 2 =	
6.	1 + 5 =		28.	10 – 8 =	
7.	6 – 1 =		29.	4 + 3 =	
8.	6 – 5 =		30.	3 + 4 =	
9.	9 + 1 =		31.	7 – 3 =	
10.	1 + 9 =		32.	7 – 4 =	
11.	10 – 1 =		33.	5 + 3 =	
12.	10 – 9 =		34.	3 + 5 =	
13.	4 + 2 =		35.	8 – 3 =	
14.	2 + 4 =		36.	8 – 5 =	
15.	6 – 2 =		37.	7 + 3 =	
16.	6 – 4 =		38.	3 + 7 =	
17.	7 + 2 =		39.	10 – 3 =	
18.	2 + 7 =		40.	10 – 7 =	
19.	9 – 2 =		41.	5 + 4 =	
20.	9 – 7 =		42.	4 + 5 =	
21.	5 + 2 =		43.	9 – 5 =	
22.	2 + 5 =		44.	9 – 4 =	

Lesson 11:   Count the total value of ones, tens, and hundreds with place value disks.

EUREKA MATH™

©2015 Great Minds. eureka-math.org
G2-M3-TE-B2-1.3.1-01.2016

Name _____   Date _____

1. Model the numbers on your place value chart using the fewest number of blocks or disks possible.

   Partner A, use base ten blocks.
   Partner B, use place value disks.
   Compare the way your numbers look.
   Whisper the numbers in standard form and unit form.

   a. 12

   b. 124

   c. 104

   d. 299

   e. 200

2. Take turns using the place value disks to model the following numbers using the fewest place value disks possible. Whisper the numbers in standard form and unit form.

   a. 25   f. 36

   b. 250   g. 360

   c. 520   h. 630

   d. 502   i. 603

   e. 205   j. 306

©2015 Great Minds. eureka-math.org
G2-M3-TE-B2-1.3.1-01.2016

Name _____ Date _____

1. Tell the value of the following numbers.

a.

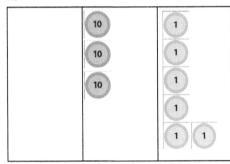

b.

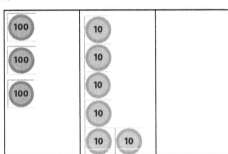

a. _____

b. _____

2. Fill in the sentences below to tell about the change from 36 to 360.

a. I changed _____ to _____.

b. I changed _____ to _____.

Lesson 11: Count the total value of ones, tens, and hundreds with place value disks.

EUREKA MATH™

©2015 Great Minds. eureka-math.org
G2-M3-TE-B2-1.3.1-01.2016

Name _____   Date _____

1. Model the following numbers for your parent using the fewest disks possible. Whisper the numbers in standard form and unit form (1 hundred 3 tens 4 ones).

    a. 15

    b. 152

    c. 102

    d. 290

    e. 300

2. Model the following numbers using the fewest place value disks possible. Whisper the numbers in standard form and unit form.

    a. 42

    b. 420

    c. 320

    d. 402

    e. 442

    f. 53

    g. 530

    h. 520

    i. 503

    j. 55

EUREKA
MATH™

Lesson 11:   Count the total value of ones, tens, and hundreds with place value disks.

167

©2015 Great Minds. eureka-math.org
G2-M3-TE-B2-1.3.1-01.2016

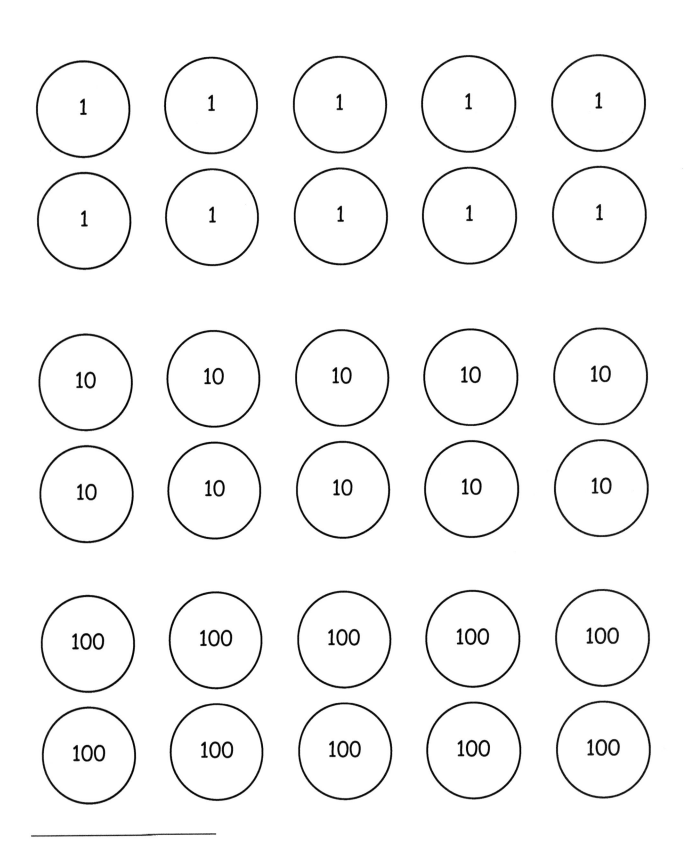

place value disks

Lesson 11: Count the total value of ones, tens, and hundreds with place value disks.

EUREKA MATH

Lesson 12

Objective: Change 10 ones for 1 ten, 10 tens for 1 hundred, and 10 hundreds for 1 thousand.

Suggested Lesson Structure

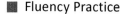

■ Fluency Practice   (10 minutes)
▨ Application Problem   (10 minutes)
▨ Concept Development   (30 minutes)
■ Student Debrief   (10 minutes)

   Total Time   **(60 minutes)**

Fluency Practice (10 minutes)

- 10 More/10 Less **2.NBT.2**   (2 minutes)
- Sprint: Sums to 10 with Teen Numbers **2.OA.2**   (8 minutes)

10 More/10 Less (2 minutes)

T: I'll say a number. You say the number that is 10 more. Wait for my signal. Ready?
T: 50.
S: 60.
T: 90.
S: 100.
T: 130.
S: 140.

Continue with 10 more, then switch to 10 less.

Sprint: Sums to 10 with Teen Numbers (8 minutes)

Materials: (S) Sums to 10 with Teen Numbers Sprint

Application Problem (10 minutes)

How many packages of 10 cookies can Collette make using 124 cookies? How many cookies does she need to complete another package of 10?

T: Let's read this problem together.

T: Visualize. Close your eyes, and see the number 124 in the different ways we've learned to represent numbers.

T: Discuss how you could solve this problem with your partner. Then, draw a model and solve.

T: (Allow two or three minutes.) Who would like to share their thinking?

S: I drew place value disks to show 124. Then, I changed the 100 disk for 10 tens, and I saw that 10 tens and 2 tens make 12 tens. Then, I drew 6 more ones disks to make another package of 10. → I knew that 100 is 10 tens and 20 is 2 tens, so I drew 12 tens. And, she needs 6 more cookies to make another ten. → I remember that 120 is 12 tens, so that's the answer. And, 6 ones plus 4 ones equals another ten.

T: Excellent reasoning! So, how many packages of 10 cookies can Collette make?

S: She can make 12 packages of 10 cookies. (Write the statement on the board.)

T: As I walked around, I noticed that most of you drew place value disks. Is it easier to draw place value disks than bundles?

S: Yes.

T: Why?

S: It's faster!

T: Yes. We want to be efficient.

T: Please add the statement to your paper if you haven't already.

Some children quickly see that there are 12 tens and 4 ones in the number 124. In this instance, adjust the number or the task to create a challenge for students working above grade level. Below are suggestions for extending the problem:

- How many packages of 10 cookies can Collette make using 124 cookies? How many cookies does she need to complete 3 more packages of 10? How many cookies will she have then?

- How many packages of 10 cookies can Collette make using 124 cookies? How many more cookies does she need to make 20 packages?

124
/ | \
(100) (20) (4)

10 tens + 2 tens = 12 tens

Collette can make 12 packages of 10 cookies.
She needs 6 more cookies to complete another package of 10.

100 + 20 + 4   = 124

120 = 12 tens

Collette can make 12 packages of 10 cookies.

4 + 6 = 10
She needs 6 cookies to make another package.

EUREKA
MATH™

Concept Development (30 minutes)

Materials: (S) Place value disks (10 ones, 10 tens, 10 hundreds), unlabeled hundreds place value chart (Lesson 8 Template) per pair

Part A: Show the Equivalence of 10 Ones and 1 Ten, 10 Tens and 1 Hundred, 10 Hundreds and 1 Thousand

Students work in pairs.

T: Slide the place value chart inside your personal white boards.

T: Show me 10 ones in two vertical columns of 5, the ten-frame way, on your place value chart.

S: (Work.)

T: What is the value of your 10 ones?

S: 10.

T: 10 potatoes?

S: 10 ones.

T: Can you change 10 ones to make a larger unit?

S: Yes.

T: What unit can you make?

S: A ten.

T: Change 10 ones for 1 ten. Did you put your 1 ten to the left or to the right?

S: To the left!

T: Yes, on the place value chart our numbers get bigger to the left!

T: Skip-count by tens on your place value chart until you have placed 10 tens.

T: Can you change to make a larger unit? (Repeat the cycle with 10 tens and 10 hundreds.)

T: Just like with our bundles, bills, and blocks, disks allow us to see how numbers work.

Part B: Count by Ones from 186 to 300 Using Place Value Disks

T: Show (silently write 186 on the board) with your place value disks. Make sure you show your units the ten-frame way.

S: (Show.)

T: Let's count up to 300 by ones. How many more ones do I need to make ten?

S: 4 ones.

T: It is easy to see because of the ten-frame format in which you have laid out your disks. Use that structure as you count to 300, please.

T: Let me hear you whisper count as you count by ones.

S: (Whisper.) 187, 188, 189, 190.

T: Pause. Can you change for a larger unit?

S: Yes. We can change 10 ones for 1 ten.

T: Do that and then keep counting with your partner up to 300. If you finish before your classmates, count down from 300 to 275.

Lesson 12: Change 10 ones for 1 ten, 10 tens for 1 hundred, and 10 hundreds for 1 thousand.

171

While students are counting, circulate and say, "Pause a moment. What number are you on? Did you just make a unit? How many more do you need to count to make the next larger unit?"

T: (Continue once most students have finished.) What were some numbers where you had to change 10 smaller units for 1 of the next unit to the left?

S: 190, 200, 250, 300, etc.

T: Use your words to tell your partner what happened when you got to both 200 and 300.

S: We made 1 ten. → We made 1 hundred. → We changed to make a ten from the ten ones. Then, that ten meant we could change 10 tens for 1 hundred.

T: Mark is expressing the change from 299 to 300 very well. Mark, will you share?

S: We changed to make a ten from the ten ones. Then, that ten meant we could change 10 tens for 1 hundred.

T: Restate Mark's explanation to your partner. You certainly may use your own words to express the same idea. (Pause while students talk.)

T: Think about the number 257. Do you remember what it looks like with your disks?

S: Yes!

T: How many more ones did 257 need to make a ten?

S: 3 ones.

T: The place value disks help us to visualize that because we put them in rows. We can easily see that we are missing 3 ones.

T: Next, you are going to count from 582 to 700, and as you go, think about how many more you need to make the next unit.

Problem Set (10 minutes)

Materials: (S) Problem Set, place value disks, unlabeled hundreds place value chart (Lesson 8 Template)

Students should do their personal best to complete the Problem Set within the allotted 10 minutes. For some classes, it may be appropriate to modify the assignment by specifying which problems they work on first. Students should solve these problems using the RDW approach used for Application Problems.

Name Jamie   Date

Count from 582 to 700 using place value disks. Change for a larger unit when necessary.

When you counted from 582 to 700:

Did you make a larger unit at..	Yes, I changed to make:		No, I need _____
1. 590?	(1 ten)	1 hundred	____ ones. ____ tens.
2. 600?	1 ten	(1 hundred)	____ ones. ____ tens.
3. 618?	1 ten	1 hundred	2 ones. ____ tens.
4. 640?	(1 ten)	1 hundred	____ ones. ____ tens.
5. 652?	1 ten	1 hundred	8 ones. ____ tens.
6. 700?	1 ten	(1 hundred)	____ ones. ____ tens.

Lesson 12:   Change 10 ones for 1 ten, 10 tens for 1 hundred, and 10 hundreds for 1 thousand.

EUREKA MATH

©2015 Great Minds. eureka-math.org
G2-M3-TE-B2-1.3.1-01.2016

Directions: Count by ones from 582 to 700 using your place value disks.

1. Model 582 with your place value disks. Count up by ones to 700.

2. Pause at each number listed on your Problem Set. At that number, did you make a larger unit?

3. If the answer is yes, tell what unit or units you made.

4. If the answer is no, tell how much more you need to make the next largest unit.

5. If you finish before time is up, model counting down to each number on the Problem Set, beginning with 700.

Student Debrief (10 minutes)

Lesson Objective: Change 10 ones for 1 ten, 10 tens for 1 hundred, and 10 hundreds for 1 thousand.

The Student Debrief is intended to invite reflection and active processing of the total lesson experience.

Invite students to review their solutions for the Problem Set. They should check work by comparing answers with a partner before going over answers as a class. Look for misconceptions or misunderstandings that can be addressed in the Debrief. Guide students in a conversation to debrief the Problem Set and process the lesson.

T: Think about the number 582. Do you remember what it looks like with your disks?

S: Yes!

T: How many more ones did 582 need to make a ten?

S: 8 ones.

T: The place value disks help us to visualize. We can easily see the 8 missing ones. Go over the answers on **MP.7** your Problem Set with a partner.

S: (Share answers.)

T: At which numbers did you not make a change?

S: 618 and 652.

T: And at which numbers did you make a change?

S: 590, 600, 640, and 700.

T: How many tens does 590 need to change 10 tens for 1 hundred?

S: 1 ten.

T: How many hundreds does 600 need to change 10 hundreds for 1 thousand?

S: 4 hundreds.

T: How many tens does 640 need to change 10 tens for 1 hundred?

T: 6 tens.

T: How many hundreds does 700 need to change 10 hundreds for 1 thousand?

S: 3 hundreds.

NOTES ON
MULTIPLE MEANS
OF ENGAGEMENT:

It may be challenging for some English language learners to say the names of larger numbers. Invite students to use their personal white boards to write each number as they count. Writing and seeing the number supports oral language development.

Lesson 12:   Change 10 ones for 1 ten, 10 tens for 1 hundred, and 10 hundreds for 1 thousand.

173

T: With your partner, count without disks from each of the numbers on the Problem Set to 900 using ones, tens, and hundreds. Remember how we used to count bundles by counting ones to complete a ten, then counting tens to complete a hundred, and then counting up by hundreds? Visualize the disks to help you.

MP.7

S: (590, 600, 700, 800, 900, etc.)

T: Today, we focused on changing 10 ones for 1 ten, 10 tens for 1 hundred, and 10 hundreds for 1 thousand.

Exit Ticket (3 minutes)

After the Student Debrief, instruct students to complete the Exit Ticket. A review of their work will help with assessing students' understanding of the concepts that were presented in today's lesson and planning more effectively for future lessons. The questions may be read aloud to the students.

A

Number Correct: _____

Sums to 10 with Teen Numbers

1.	3 + 1 =	
2.	13 + 1 =	
3.	5 + 1 =	
4.	15 + 1 =	
5.	7 + 1 =	
6.	17 + 1 =	
7.	4 + 2 =	
8.	14 + 2 =	
9.	6 + 2 =	
10.	16 + 2 =	
11.	8 + 2 =	
12.	18 + 2 =	
13.	4 + 3 =	
14.	14 + 3 =	
15.	6 + 3 =	
16.	16 + 3 =	
17.	5 + 5 =	
18.	15 + 5 =	
19.	7 + 3 =	
20.	17 + 3 =	
21.	6 + 4 =	
22.	16 + 4 =	

23.	4 + 5 =	
24.	14 + 5 =	
25.	2 + 5 =	
26.	12 + 5 =	
27.	5 + 4 =	
28.	15 + 4 =	
29.	3 + 4 =	
30.	13 + 4 =	
31.	3 + 6 =	
32.	13 + 6 =	
33.	7 + 1 =	
34.	17 + 1 =	
35.	8 + 1 =	
36.	18 + 1 =	
37.	4 + 3 =	
38.	14 + 3 =	
39.	4 + 1 =	
40.	14 + 1 =	
41.	5 + 3 =	
42.	15 + 3 =	
43.	4 + 4 =	
44.	14 + 4 =	

EUREKA MATH

Lesson 12: Change 10 ones for 1 ten, 10 tens for 1 hundred, and 10 hundreds for 1 thousand.

175

B

Number Correct: _____

Improvement: _____

Sums to 10 with Teen Numbers

1.	2 + 1 =	
2.	12 + 1 =	
3.	4 + 1 =	
4.	14 + 1 =	
5.	6 + 1 =	
6.	16 + 1 =	
7.	3 + 2 =	
8.	13 + 2 =	
9.	5 + 2 =	
10.	15 + 2 =	
11.	7 + 2 =	
12.	17 + 2 =	
13.	5 + 3 =	
14.	15 + 3 =	
15.	7 + 3 =	
16.	17 + 3 =	
17.	6 + 3 =	
18.	16 + 3 =	
19.	5 + 4 =	
20.	15 + 4 =	
21.	1 + 9 =	
22.	11 + 9 =	

23.	9 + 1 =	
24.	19 + 1 =	
25.	5 + 1 =	
26.	15 + 1 =	
27.	5 + 3 =	
28.	15 + 3 =	
29.	6 + 2 =	
30.	16 + 2 =	
31.	3 + 6 =	
32.	13 + 6 =	
33.	7 + 2 =	
34.	17 + 2 =	
35.	1 + 8 =	
36.	11 + 8 =	
37.	3 + 5 =	
38.	13 + 5 =	
39.	4 + 2 =	
40.	14 + 2 =	
41.	5 + 4 =	
42.	15 + 4 =	
43.	1 + 6 =	
44.	11 + 6 =	

Lesson 12:   Change 10 ones for 1 ten, 10 tens for 1 hundred, and 10 hundreds for 1 thousand.

EUREKA MATH™

Name _____   Date _____

Count from **582 to 700** using place value disks. Change for a larger unit when necessary.

When you counted from **582 to 700**:

Did you make a larger unit at...	Yes, I changed to make:		No, I need _____
1. 590?	1 ten	1 hundred	____ ones. ____ tens.
2. 600?	1 ten	1 hundred	____ ones. ____ tens.
3. 618?	1 ten	1 hundred	____ ones. ____ tens.
4. 640?	1 ten	1 hundred	____ ones. ____ tens.
5. 652?	1 ten	1 hundred	____ ones. ____ tens.
6. 700?	1 ten	1 hundred	____ ones. ____ tens.

Lesson 12:   Change 10 ones for 1 ten, 10 tens for 1 hundred, and 10 hundreds for 1 thousand.

177

Name _____ Date _____

1. Match to show the equivalent value.

    a. 10 ones   1 hundred

    b. 10 tens   1 thousand

    c. 10 hundreds   1 ten

2. Draw disks on the place value chart to show 348.

    a. How many more ones to make a ten?   _____ ones

    b. How many more tens to make a hundred?   _____ tens

    c. How many more hundreds to make a thousand? _____ hundreds

178   Lesson 12:   Change 10 ones for 1 ten, 10 tens for 1 hundred, and 10 hundreds for 1 thousand.

EUREKA MATH

Name _____   Date _____

Count by ones from **368 to 500**. Change for a larger unit when necessary.

When you counted from **368 to 500**:

Did you make a larger unit at...	Yes, I changed to make:		No, I need _____
1. 377?	1 ten	1 hundred	____ ones. ____ tens.
2. 392?	1 ten	1 hundred	____ ones. ____ tens.
3. 400?	1 ten	1 hundred	____ ones. ____ tens.
4. 418?	1 ten	1 hundred	____ ones. ____ tens.
5. 463?	1 ten	1 hundred	____ ones. ____ tens.
6. 470?	1 ten	1 hundred	____ ones. ____ tens.

Lesson 12: Change 10 ones for 1 ten, 10 tens for 1 hundred, and 10 hundreds for 1 thousand.

179

©2015 Great Minds. eureka-math.org
G2-M3-TE-B2-1.3.1-01.2016

Lesson 13

Objective: Read and write numbers within 1,000 after modeling with place value disks.

Suggested Lesson Structure

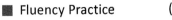

■ Fluency Practice   (10 minutes)
▨ Application Problem   (10 minutes)
▨ Concept Development   (30 minutes)
▨ Student Debrief   (10 minutes)

   Total Time   **(60 minutes)**

Fluency Practice (10 minutes)

- Sprint: Place Value Counting to 100 **2.NBT.3**   (8 minutes)
- 100 More/100 Less **2.NBT.2**   (1 minute)
- How Many Tens/How Many Hundreds **2.NBT.1**   (1 minute)

Sprint: Place Value Counting to 100 (8 minutes)

Materials: (S) Place Value Counting to 100 Sprint

100 More/100 Less (1 minute)

T: I'll say a number. You say the number that is 100 more. Wait for my signal. Ready?
T: 70.
S: 170.
T: 200.
S: 300.
T: 480.
S: 580.
T: 900.
S: 1,000.

Continue with 100 more, and then switch to 100 less.

Lesson 13:   Read and write numbers within 1,000 after modeling with place value
                        disks.

©2015 Great Minds. eureka-math.org
G2-M3-TE-B2-1.3.1-01.2016

How Many Tens/How Many Hundreds (1 minute)

T: I'll say a number. You say how many tens are in that number. For example, I say, "14 ones."
     You say, "1 ten." Wait for my signal. Ready?

T: 20 ones.

S: 2 tens!

T: 28 ones.

S: 2 tens!

T: 64 ones.

S: 6 tens!

T: 99 ones.

S: 9 tens!

Continue in this manner, and then switch to asking how many hundreds.

T: 15 tens.

S: 1 hundred!

T: 29 tens.

S: 2 hundreds!

T: 78 tens.

S: 7 hundreds!

Application Problem (10 minutes)

Sarah's mom bought 4 boxes of crackers. Each box had 3 smaller packs of 10 inside. How many crackers
were in the 4 boxes?

T: Read this problem with me.

T: We always have to pay special attention to the
     information given.

T: How many boxes are there?

S: 4.

T: What is inside each box?

S: 3 packs of 10 crackers.

T: What unit are we solving for, boxes or crackers?
     Reread the question, and then tell your partner.

S: Crackers.

T: Correct. Now, discuss with your partner what you
     could draw that would help you answer the question.

NOTES ON
MULTIPLE MEANS
OF REPRESENTATION:

The difference between the words
packs and *boxes* could be difficult for
English language learners to
understand. Before beginning, it may
be helpful to define these words with a
simple labeled sketch that students can
refer to as they read the problem.

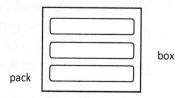

S: I drew 4 boxes and wrote 10, 10, 10 in each one.
Then, I skip-counted by tens and got 120. → I drew
4 big circles and put 3 ten disks inside each. Then, I
used doubles. 3 tens + 3 tens is 6 tens, and 6 tens +
6 tens is 12 tens or 120. → I drew the same picture as
Yesenia, but I skip-counted 3, 6, 9, 12. And, since
they're tens, I said 30, 60, 90, 120.

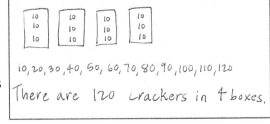

10, 20, 30, 40, 50, 60, 70, 80, 90, 100, 110, 120

There are 120 crackers in 4 boxes.

T: Great strategies for solving! So, what is the
answer to the question?

S: There are 120 crackers in the 4 boxes.

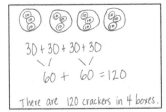

30 + 30 + 30 + 30
60 + 60 = 120
There are 120 crackers in 4 boxes.

Concept Development (30 minutes)

Drawing Place Value Disks to Represent Numbers (10 minutes)

Materials: (T) Plenty of white board space, empty number line (Template) (S) Personal white board, empty
number line (Template)

T: I'm going to draw some pictures of numbers. As I draw, count out loud for me.

T: (Draw an unlabeled place value chart on the board, and draw pictures of the disks to represent 322.)

S: 100, 200, 300, 310, 320, 321, 322.

T: What is the value of the number on my place value chart? Write the value on your personal white
board. Show the value to me at the signal.

S: (Show 322.)

T: Excellent. Try another. (Silently draw disks to represent 103 as students count the value.)

S: 100, 101, 102, 103.

T: What's the total value of this new number? Write it on your personal board. Show the value to me
at the signal.

S: (Show 103.)

T: Now, we'll try a new process. I'm thinking of a number. Don't count while I draw. Wait until I have
finished drawing before you whisper its value to your partner.

T: (Silently and quickly draw 281 on the place value chart. Be sure to draw the ten-frame way, as
modeled in the Problem Set below.)

T: Write this new number on your personal board.

T: Here is another one. (A possible sequence would be 129, 710, 807, 564.)

T: What is it about the way I am drawing that is making it easy for you to tell the value of my number so
quickly? Talk to your partner.

S: The labels are easy to read. → She draws the place value disks like a ten-frame. → The place value
chart makes the units easy to see.

Lesson 13:   Read and write numbers within 1,000 after modeling with place value
disks.

T: I hear a lot of interesting ideas. We have some great tools here. What tools are we using?

S: A place value chart. → Place value disks.
     → The ten-frame.

T: Now, it is your turn to represent some numbers by drawing place value disks.

Problem Set (10 minutes)

NOTES ON
MULTIPLE MEANS
OF ACTION AND
EXPRESSION:

Some students may need to use place value disks to model numbers before drawing them. Allow students to use the disks for the first problem, but wean them off as quickly as possible. For each digit, prompt students to visualize the disks that they will draw, say the number, then make the drawing before going on to the next digit. This helps build confidence by creating manageable steps.

Students should do their personal best to complete the Problem Set within the allotted 10 minutes. For some classes, it may be appropriate to modify the assignment by specifying which problems they work on first. Some problems do not specify a method for solving. Students should solve these problems using the RDW approach used for Application Problems.

Directions: Draw the numbers indicated using place value disks drawn the ten-frame way.

Notes on drawing place value disks:

- Have the students draw the value of the unit first, and then circle it. (Otherwise, students might draw the circle first and then cram the unit's value inside.)

- Have the students start drawing from the top down in each place of the place value chart, filling their column of 5 (if the number is 5 or greater).

- Start from the bottom up to build toward the other five for 6, 7, 8, and 9.

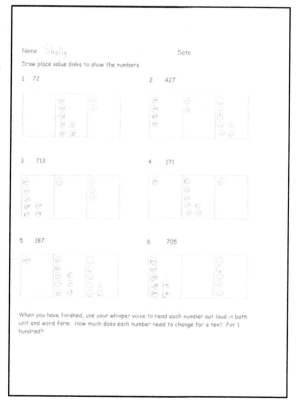

Estimating Numbers on the Empty Number Line (10 minutes)

T: Let's represent the same numbers from our Problem Set on empty number lines. Imagine we are traveling from 0 to 72.

T: (Point to the beginning and end of the first number line.) Here is 0's address for now. And here is 72's address at the other end of the number line.

T: How many tens am I going to travel?

S: 7 tens.

T: I would like the 7 jumps to be as equal as I can make them. I like drawing little arrows to show the jumps I make. Count for me. (Draw 7 large jumps and 2 small jumps.)

S: 1 ten, 2 tens, 3 tens, 4 tens, 5 tens, 6 tens, 7 tens, 1 one, 2 ones.

        Lesson 13:   Read and write numbers within 1,000 after modeling with place value   183
                                              disks.

©2015 Great Minds. eureka-math.org
G2-M3-TE-B2-1.3.1-01.2016

T: Below, I'm going to draw my disks (as pictured).

T: Now, you try. Here is an empty number line template. Use a pencil because you might erase a few times. Make your address for 0 and 72, and then get to 72 the best you can with 7 tens and 2 ones.

T: (Circulate and support. Move them on to 427. "What units do you have in that number?" "Which is the largest?" "Draw the disks below to show the units within each hop." Continue with the following possible sequence: 713, 171, 187.)

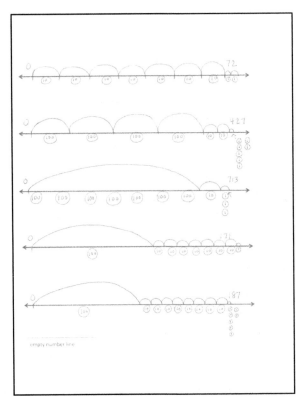

Encourage students to have fun and think about the best way to show each number on the empty number line. Accept all reasonable work. Do not be overly prescriptive. Watch for students who make different units the same size. "Yes, the disks are the same size, but will the hops be the same size on the number line?" This is an estimation exercise and a chance to consider the size of a unit while working with the disks.

Student Debrief (10 minutes)

Lesson Objective: Read and write numbers within 1,000 after modeling with place value disks.

Materials: (T) Base ten bundles of straws on the carpet, Problem Sets

The Student Debrief is intended to invite reflection and active processing of the total lesson experience.

Invite students to review their solutions for the Problem Set. They should check work by comparing answers with a partner before going over answers as a class. Look for misconceptions or misunderstandings that can be addressed in the Debrief. Guide students in a conversation to debrief the Problem Set and process the lesson.

T: Bring your work to the carpet. Check your partner's place value charts. Make sure the correct number of units is drawn for each one and that they are easy to read. Make sure they are in the correct place, too.

MP.6

S: (Share.)

T: Let's start by analyzing our place value charts. In each number there is a 7. With your partner, review the values of the sevens. (Allow time for partners to review.)

T: Read the numbers in order from Problems 1 through 6 when I give the signal.

S: (Read.)

T: Discuss with your partner the bundles that would match each of your six numbers. I have the bundles on the carpet here for you to refer to.

©2015 Great Minds. eureka-math.org
G2-M3-TE-B2-1.3.1-01.2016

T: Now, share your number lines with your partner. Explain your thinking about the size of your hops.

S: I knew I had to get in 7 hops by the end of the line, so I made them smaller here than in this one. → It's interesting because this line was 427, and this line was 72. → So, on this one, I made 4 really big hops, 2 small ones, and then 7 minis. → Then, 700. I decided to just make one big hop for all the hundreds.

T: Let's read through the numbers we showed both on the place value chart and on the empty number line.

S: 72, 427, 713, 171, 187.

T: As we already saw, each of our numbers has a 7 in it. Show your partner how you represented the **MP.6** 7 in each number on your number line. Why are they different?

S: This was 7 hundreds, and this was 7 ones. So, these were little, and these were big. → Both of these numbers had 7 in the tens place, but 72 is smaller than 171, so the hops were bigger when I was only going to 72. → 705 and 713 both have 7 hundreds. On this number line, I hopped 7 times, but on this number line I made one big jump for all seven. I guess that it's just about the same, though.

T: So, I'm hearing you say that the biggest difference was in the way the 7 tens in 171 and 72 looked, the unit in the middle.

T: It's so interesting because a number could be counting something really small like 70 grains of rice or something really big like 70 planets! We read and write numbers, and they describe things. Turn and talk to your partner. What could our number 427 be describing?

S: 427 apples. → 427 students. → 427 ants. → 427 stars.

Exit Ticket (3 minutes)

After the Student Debrief, instruct students to complete the Exit Ticket. A review of their work will help with assessing students' understanding of the concepts that were presented in today's lesson and planning more effectively for future lessons. The questions may be read aloud to the students.

Lesson 13:   Read and write numbers within 1,000 after modeling with place value disks.

185

A

Number Correct: _____

Place Value Counting to 100

1.	5 tens	
2.	6 tens 2 ones	
3.	6 tens 3 ones	
4.	6 tens 8 ones	
5.	60 + 4 =	
6.	4 + 60 =	
7.	8 tens	
8.	9 tens 4 ones	
9.	9 tens 5 ones	
10.	9 tens 8 ones	
11.	90 + 6 =	
12.	6 + 90 =	
13.	6 tens	
14.	7 tens 6 ones	
15.	7 tens 7 ones	
16.	7 tens 3 ones	
17.	70 + 8 =	
18.	8 + 70 =	
19.	9 tens	
20.	8 tens 1 one	
21.	8 tens 2 ones	
22.	8 tens 7 ones	

23.	80 + 4 =	
24.	4 + 80 =	
25.	7 tens	
26.	5 tens 8 ones	
27.	5 tens 9 ones	
28.	5 tens 2 ones	
29.	50 + 7 =	
30.	7 + 50 =	
31.	10 tens	
32.	7 tens 4 ones	
33.	80 + 3 =	
34.	7 + 90 =	
35.	6 tens + 10 =	
36.	9 tens 3 ones	
37.	70 + 2 =	
38.	3 + 50 =	
39.	60 + 2 tens =	
40.	8 tens 6 ones	
41.	90 + 2 =	
42.	5 + 60 =	
43.	8 tens 20 ones	
44.	30 + 7 tens =	

Lesson 13: Read and write numbers within 1,000 after modeling with place value disks.

EUREKA MATH

B

Place Value Counting to 100

1.	6 tens	
2.	5 tens 2 ones	
3.	5 tens 3 ones	
4.	5 tens 8 ones	
5.	4 + 60 =	
6.	50 + 4 =	
7.	4 + 50 =	
8.	8 tens 4 ones	
9.	8 tens 5 ones	
10.	8 tens 8 ones	
11.	80 + 6 =	
12.	6 + 80 =	
13.	7 tens	
14.	9 tens 6 ones	
15.	9 tens 7 ones	
16.	9 tens 3 ones	
17.	90 + 8 =	
18.	8 + 90 =	
19.	5 tens	
20.	6 tens 1 one	
21.	6 tens 2 ones	
22.	6 tens 7 ones	

23.	60 + 4 =	
24.	4 + 60 =	
25.	8 tens	
26.	7 tens 8 ones	
27.	7 tens 9 ones	
28.	7 tens 2 ones	
29.	70 + 5 =	
30.	5 + 70 =	
31.	10 tens	
32.	5 tens 6 ones	
33.	60 + 3 =	
34.	6 + 70 =	
35.	5 tens + 10 =	
36.	7 tens 4 ones	
37.	80 + 3 =	
38.	2 + 90 =	
39.	70 + 2 tens	
40.	6 tens 8 ones	
41.	70 + 3 =	
42.	7 + 80 =	
43.	9 tens 10 ones	
44.	40 + 6 tens =	

Lesson 13: Read and write numbers within 1,000 after modeling with place value disks.

187

Name _____   Date _____

Draw place value disks to show the numbers.

1. 72

2. 427

3. 713

4. 171

5. 187

6. 705

When you have finished, use your whisper voice to read each number out loud in both unit and word form. How much does each number need to change for a ten?
For 1 hundred?

Lesson 13:   Read and write numbers within 1,000 after modeling with place value disks.

EUREKA
MATH™

Name _____ Date _____

1. Draw place value disks to show the numbers.

   a. 560   b. 506

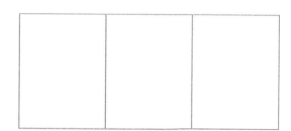

2. Draw and label the jumps on the number line to move from 0 to 141.

EUREKA
MATH™

Lesson 13:   Read and write numbers within 1,000 after modeling with place value
disks.

189

©2015 Great Minds. eureka-math.org
G2-M3-TE-B2-1.3.1-01.2016

Name _____   Date _____

Draw place value disks to show the numbers.

1. 43

2. 430

3. 270

4. 720

5. 702

6. 936

When you have finished, use your whisper voice to read each number out loud in both unit and word form. How much does each number need to change for a ten?
For 1 hundred?

Lesson 13:   Read and write numbers within 1,000 after modeling with place value disks.

EUREKA
MATH

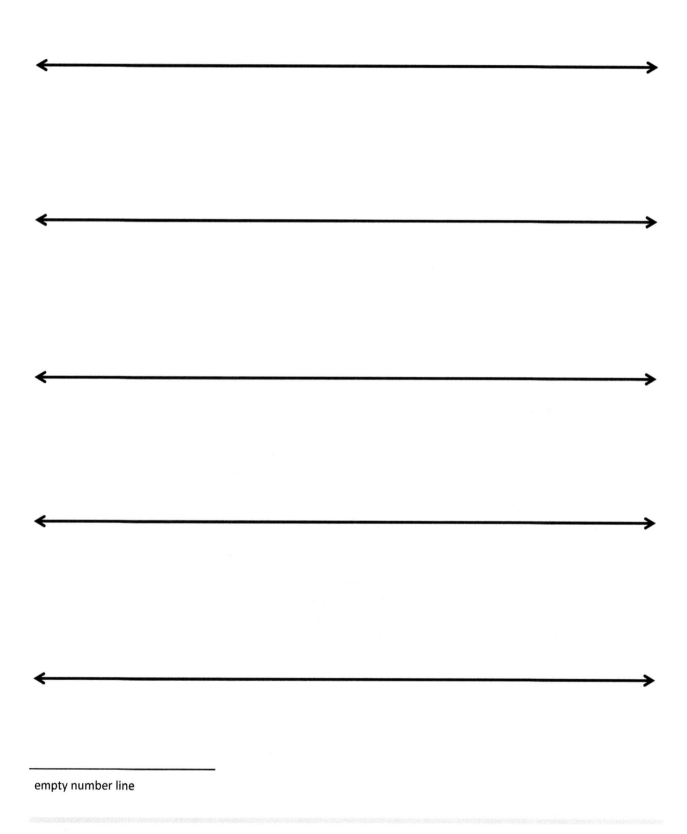

empty number line

Lesson 13:   Read and write numbers within 1,000 after modeling with place value
disks.

191

©2015 Great Minds. eureka-math.org
G2-M3-TE-B2-1.3.1-01.2016

Lesson 14

Objective: Model numbers with more than 9 ones or 9 tens; write in expanded, unit, standard, and word forms.

Suggested Lesson Structure

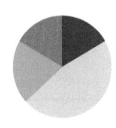

■ Fluency Practice   (10 minutes)
▨ Application Problem   (12 minutes)
▢ Concept Development   (28 minutes)
■ Student Debrief   (10 minutes)

   Total Time   **(60 minutes)**

Fluency Practice (10 minutes)

▪ Sprint: Review of Subtraction in the Teens **2.OA.2**   (8 minutes)
▪ Happy Counting Up and Down by Ones Crossing 100 **2.NBT.2**   (2 minutes)

Sprint: Review of Subtraction in the Teens (8 minutes)

Materials   (S) Review of Subtraction in the Teens Sprint

Happy Counting Up and Down by Ones Crossing 100 (2 minutes)

   T: Let's play Happy Counting!
   T: Watch my fingers to know whether to count up or down. A closed hand means stop. (Show signals while explaining.)
   T: We'll count by ones, starting at 76. Ready? (Rhythmically point up until a change is desired. Show a closed hand, and then point down. Continue, mixing it up.)
   S: 76, 77, 78, 79, 80, 81. (Switch direction.) 80, 79, 78. (Switch direction.) 79, 80, 81, 82, 83, 84, 85, 86, 87, 88, 89, 90, 91, 92. (Switch direction.) 91, 90, 89, 88, 87. (Switch direction.) 88, 89, 90, 91, 92, 93, 94, 95. (Switch direction.) 94, 93. (Switch direction.) 94, 95, 96, 97, 98, 99, 100, 101, 102, 103. (Switch direction.) 102, 101, 100, 99, 98. (Switch direction.) 99, 100, 101, 102, 103, 104, 105, 106.

Application Problem (12 minutes)

A second grade class has 23 students. What is the total number of fingers of all the students?

T: Read this problem with me.

T: I'm very curious to see what you'll draw to solve this! Talk with your partner to share ideas, and then I'll give you two minutes to draw and label your picture.

T: (Allow several minutes.) Who would like to share their thinking?

S: I drew 23 circles to be the 23 students. Then, I put the number 10 in each to be the 10 fingers for everybody. Then, I skip-counted by tens and got to 230. → I drew 23 ten disks because each student has 10 fingers. Then, I circled 1 group of 10 circles and wrote 100 because 10 tens equals 100. Then, I circled another group of 10 circles. That made 200. And, there were 3 tens left, which is 30. So, the answer is 230.

T: 230 what?

S: 230 fingers!

T: Why is it easier to draw 23 ten disks than, say, 23 sets of hands?

S: It's faster! → It takes longer to draw two hands for every student instead of just 1 circle for each student.

T: Good reasoning! It's good to be fast if you can be accurate, but it's also important to use a strategy that makes sense to you.

T: So, how many fingers do 23 students have?

S: 23 students have 230 fingers!

T: Please add that statement to your paper.

Concept Development (28 minutes)

Materials: (S) Place value disks (9 hundreds, 15 tens, 15 ones), unlabeled hundreds place value chart (Lesson 8 Template)

T: Slide the place value chart inside your personal white boards.

T: On your place value chart, show me the number 14.

S: (Show.)

T: What disks did you use from greatest to least?

NOTES ON
MULTIPLE MEANS
OF ACTION AND
EXPRESSION:

Adjust the number or the task to challenge students working above grade level. Below are two suggestions for extending the problem:

- A second-grade class has 23 students. What is the total number of fingers of the students? What is the total number of toes? How many fingers and toes are there altogether?

- A second-grade class has 23 students. What is the total number of fingers of the students? How many more students need to join the class so that there are 300 student fingers in all?

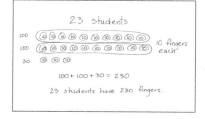

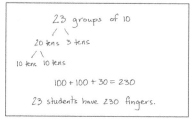

Lesson 14: Model numbers with more than 9 ones or 9 tens; write in expanded, unit, standard, and word forms.

193

©2015 Great Minds. eureka-math.org
G2-M3-TE-B2-1.3.1-01.2016

S: 1 ten and 4 ones.

T: Change 1 ten for 10 ones. (Pause as students work.) What disks did you use this time?

S: 14 ones.

T: Discuss with your partner why this statement is true. (Silently write 1 ten 2 ones = 12 ones.)

S: Yes, it is true. → It's true because 1 ten is 10 ones, and 10 + 2 is 12 ones. → Yes, but my teacher said you can't have more than 9 ones. → It's okay to use more. It's just faster to use a ten.

T: Show me the number 140 with your disks.

S: (Show.)

T: What place value disks did you use from greatest to least?

S: 1 hundred 4 tens.

T: Change 1 hundred for 10 tens. (Pause as students work.) What disks did you use this time?

S: 14 tens.

T: Touch and count by tens to find the total value of your tens.

S: 10, 20, 30, 40, 50, 60, 70, 80, 90, 100, 110, 120, 130, 140.

T: What is the value of 14 tens? Answer in a full sentence, "The value of 14 tens is…

S: The value of 14 tens is 140.

T: Discuss why this statement is true with your partner. (Silently write: 1 hundred 4 tens = 14 tens.)

T: Now, discuss with your partner why this is true. (Silently write 14 tens = 140 ones.)

T: Show me the number 512.

T: What disks did you use?

S: 5 hundreds 1 ten 2 ones.

T: Change 1 ten for 10 ones. (Pause as students work.) What disks did you use?

S: 5 hundreds 12 ones.

T: Discuss why the statement is true. (Write 5 hundreds 1 ten 2 ones = 5 hundreds 12 ones. Continue with more guided examples if necessary with a small group.)

T: Let's try some more. First, model A and then B. Tell the total value of each number you model.

A	B
1 hundred 5 tens 2 ones	15 tens 2 ones
11 tens	1 hundred 1 ten
1 ten 3 ones	13 ones
12 tens 9 ones	1 hundred 2 tens 9 ones

Lesson 14:   Model numbers with more than 9 ones or 9 tens; write in expanded, unit, standard, and word forms.

EUREKA
MATH™

©2015 Great Minds. eureka-math.org
G2-M3-TE-B2-1.3.1-01.2016

Problem Set (12 minutes)

Materials: (S) Problem Set

Students should do their personal best to complete the Problem Set within the allotted 10 minutes. For some classes, it may be appropriate to modify the assignment by specifying which problems they work on first. Some problems do not specify a method for solving. Students should solve these problems using the RDW approach used for Application Problems.

Directions: Represent each number two ways on the place value charts. The instructions will tell you what units to use.

Student Debrief (10 minutes)

Lesson Objective: Model numbers with more than 9 ones or 9 tens; write in expanded, unit, standard, and word forms.

The Student Debrief is intended to invite reflection and active processing of the total lesson experience.

Invite students to review their solutions for the Problem Set. They should check work by comparing answers with a partner before going over answers as a class. Look for misconceptions or misunderstandings that can be addressed in the Debrief. Guide students in a conversation to debrief the Problem Set and process the lesson.

T: Bring your Problem Set to our Debrief.

T: Check your work carefully with a partner. How did you show each number? I will circulate and look at your drawings, too.

T: (Allow two minutes.) Which ones were hard for you?

T: (Ask questions, especially with the third page. If no one is forthcoming, choose one many struggled with.)

T: Let's look at Problem 1(c). What number is written?

S: 206.

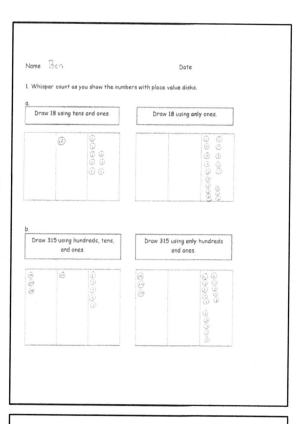

     Lesson 14:   Model numbers with more than 9 ones or 9 tens; write in expanded,   195
                               unit, standard, and word forms.

©2015 Great Minds. eureka-math.org
G2-M3-TE-B2-1.3.1-01.2016

T: Say 206 in expanded form.

S: 200 + 6.

T: 100 + 100 is…?

S: 200.

T: 100 is how many tens?

S: 10 tens.

T: 10 tens + 10 tens is…?

S: 20 tens.

T: 20 tens is…?

S: 200.

T: 206 = 2 hundreds 6 ones = 20 tens 6 ones. Talk to your partner about why this is true.

T: We can have more than 9 units. Let's try some.

T: The value of 30 tens is…?

S: 300.

T: 18 tens?

S: 180.

T: Excellent.

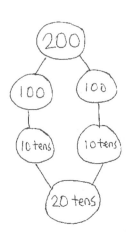

Exit Ticket (3 minutes)

After the Student Debrief, instruct students to complete the Exit Ticket. A review of their work will help with assessing students' understanding of the concepts that were presented in today's lesson and planning more effectively for future lessons. The questions may be read aloud to the students.

NOTES ON MULTIPLE MEANS OF ENGAGEMENT:

The Debrief relies heavily on oral language and automaticity with that language. If students need support, it may be appropriate to have them answer some questions on their personal white boards. Alternatively, ask students to chorally respond at a signal to allow for wait time between responses.

196   Lesson 14:   Model numbers with more than 9 ones or 9 tens; write in expanded, unit, standard, and word forms.

EUREKA MATH

©2015 Great Minds. eureka-math.org
G2-M3-TE-B2-1.3.1-01.2016

A

Number Correct: _____

Review of Subtraction in the Teens

1.	3 – 1 =		23.	7 – 4 =	
2.	13 – 1 =		24.	17 – 4 =	
3.	5 – 1 =		25.	7 – 5 =	
4.	15 – 1 =		26.	17 – 5 =	
5.	7 – 1 =		27.	9 – 5 =	
6.	17 – 1 =		28.	19 – 5 =	
7.	4 – 2 =		29.	7 – 6 =	
8.	14 – 2 =		30.	17 – 6 =	
9.	6 – 2 =		31.	9 – 6 =	
10.	16 – 2 =		32.	19 – 6 =	
11.	8 – 2 =		33.	8 – 7 =	
12.	18 – 2 =		34.	18 – 7 =	
13.	4 – 3 =		35.	9 – 8 =	
14.	14 – 3 =		36.	19 – 8 =	
15.	6 – 3 =		37.	7 – 3 =	
16.	16 – 3 =		38.	17 – 3 =	
17.	8 – 3 =		39.	5 – 4 =	
18.	18 – 3 =		40.	15 – 4 =	
19.	6 – 4 =		41.	8 – 5 =	
20.	16 – 4 =		42.	18 – 5 =	
21.	8 – 4 =		43.	8 – 6 =	
22.	18 – 4 =		44.	18 – 6 =	

EUREKA MATH

Lesson 14: Model numbers with more than 9 ones or 9 tens; write in expanded, unit, standard, and word forms.

197

©2015 Great Minds. eureka-math.org
G2-M3-TE-B2-1.3.1-01.2016

B

Number Correct: _____

Improvement: _____

Review of Subtraction in the Teens

1.	2 – 1 =		23.	9 – 4 =	
2.	12 – 1 =		24.	19 – 4 =	
3.	4 – 1 =		25.	6 – 5 =	
4.	14 – 1 =		26.	16 – 5 =	
5.	6 – 1 =		27.	8 – 5 =	
6.	16 – 1 =		28.	18 – 5 =	
7.	3 – 2 =		29.	8 – 6 =	
8.	13 – 2 =		30.	18 – 6 =	
9.	5 – 2 =		31.	9 – 6 =	
10.	15 – 2 =		32.	19 – 6 =	
11.	7 – 2 =		33.	9 – 7 =	
12.	17 – 2 =		34.	19 – 7 =	
13.	5 – 3 =		35.	9 – 8 =	
14.	15 – 3 =		36.	19 – 8 =	
15.	7 – 3 =		37.	8 – 3 =	
16.	17 – 3 =		38.	18 – 3 =	
17.	9 – 3 =		39.	6 – 4 =	
18.	19 – 3 =		40.	16 – 4 =	
19.	5 – 4 =		41.	9 – 5 =	
20.	15 – 4 =		42.	19 – 5 =	
21.	7 – 4 =		43.	7 – 6 =	
22.	17 – 4 =		44.	17 – 6 =	

Lesson 14: Model numbers with more than 9 ones or 9 tens; write in expanded, unit, standard, and word forms.

EUREKA MATH™

©2015 Great Minds. eureka-math.org
G2-M3-TE-B2-1.3.1-01.2016

Name _____ Date _____

1. Whisper count as you show the numbers with place value disks.

   a.

Draw 18 using tens and ones.

Draw 18 using **only** ones.

   b.

Draw 315 using hundreds, tens, and ones.

Draw 315 using **only** hundreds and ones.

Lesson 14: Model numbers with more than 9 ones or 9 tens; write in expanded, unit, standard, and word forms.

199

©2015 Great Minds. eureka-math.org
G2-M3-TE-B2-1.3.1-01.2016

c.

| Draw 206 using hundreds, tens, and ones. | Draw 206 using **only** tens and ones. |

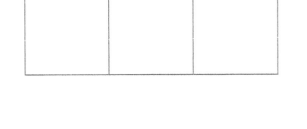

2. Whisper-talk the numbers and words as you fill in the blanks. Start by using the place value charts from Problem 1 to help you.

  a.   18 = _____ hundreds _____ tens _____ ones

       18 = _____ ones

  b. 315 = _____ hundreds _____ tens _____ ones

      315 = _____ hundreds _____ ones

  c. 206 = _____ hundreds _____ tens _____ ones

      206 = _____ tens _____ ones

  d. 419 = _____ hundreds _____ tens _____ ones

      419 = _____ tens _____ ones

Lesson 14:   Model numbers with more than 9 ones or 9 tens; write in expanded, unit, standard, and word forms.

EUREKA MATH™

e. 570 = _____ hundreds _____ tens

   570 = _____ tens

f. 748 = _____ hundreds _____ ones

   748 = _____ tens _____ ones

g. 909 = _____ hundreds _____ ones

   909 = _____ tens _____ ones

3. Mr. Hernandez's class wants to trade 400 tens rods for hundreds flats with Mr. Harrington's class. How many hundreds flats are equal to 400 tens rods?

Lesson 14:   Model numbers with more than 9 ones or 9 tens; write in expanded,   **201**
                 unit, standard, and word forms.

©2015 Great Minds. eureka-math.org
G2-M3-TE-B2-1.3.1-01.2016

Name _____   Date _____

1. Whisper count as you show the numbers with place value disks.

    a. Draw 241 using hundreds, tens, and ones.

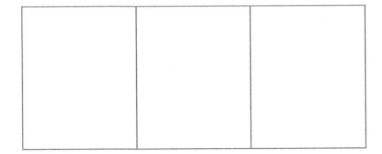

    b. Draw 241 using only tens and ones.

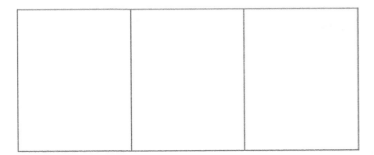

2. Fill in the blanks.

    a. 45 = _____ hundreds _____ tens _____ ones

        45 = _____ ones

    b. 682 = _____ hundreds _____ tens _____ ones

        682 = _____ hundreds _____ ones

EUREKA MATH

©2015 Great Minds. eureka-math.org
G2-M3-TE-B2-1.3.1-01.2016

Name _____   Date _____

1. Whisper-talk the numbers and words as you fill in the blanks.

a. 16 = _____ tens _____ ones

16 = _____ ones

b. 217 = _____ hundreds _____ tens _____ ones

217 = _____ hundreds _____ ones

c. 320 = _____ hundreds _____ tens _____ ones

320 = _____ tens _____ ones

d. 139 = _____ hundreds _____ tens _____ ones

139 = _____ tens _____ ones

e. 473 = _____ hundreds _____ tens _____ ones

473 = _____ tens _____ ones

f. 680 = _____ hundreds _____ tens

680 = _____ tens

g. 817 = _____ hundreds _____ ones

817 = _____ tens _____ ones

EUREKA
MATH™

Lesson 14:   Model numbers with more than 9 ones or 9 tens; write in expanded,
unit, standard, and word forms.

203

©2015 Great Minds. eureka-math.org
G2-M3-TE-B2-1.3.1-01.2016

h. 921 = _____ hundreds _____ ones

921 = _____ tens _____ ones

2. Write down how you can skip-count by ten from 350 to 240. You might use place value disks, number lines, bundles, or numbers.

Model numbers with more than 9 ones or 9 tens; write in expanded, unit, standard, and word forms.

EUREKA
MATH™

Lesson 15

Objective: Explore a situation with more than 9 groups of ten.

Suggested Lesson Structure

- Fluency Practice   (12 minutes)
- Concept Development   (30 minutes)
- Student Debrief   (18 minutes)

  Total Time   **(60 minutes)**

Fluency Practice (12 minutes)

- Sprint: Expanded Notation **2.NBT.3**   (8 minutes)
- Compare Numbers 0–99 Using <, >, = **2.NBT.4**   (4 minutes)

Sprint: Expanded Notation (8 minutes)

Materials: (S) Expanded Notation Sprint

Compare Numbers 0–99 Using <, >, = (4 minutes)

Materials: (T) 1 set of pre-cut <, >, = symbols (Template 1)
             (S) Small resealable bag containing 2 sets of pre-cut
             digit cards 0–9 (Template 2) per student, 1 set of
             pre-cut <, >, = symbol cards (Template 1) per pair

Students are seated in partners at their tables.

- T: Take the digit cards out of your small resealable bag.
       Use the cards to build a number from 0–99. Take 10
       seconds.
- T: Compare numbers with your partner. Place the
       appropriate symbol (show <, >, =) between them.
- T: Read your number sentence to your partner using the
       words *greater than, less than,* or *equal to.* Then, use
       the language of units to explain how you know the
       number sentence is true.
- T: For example, you might say, "34 is less than 67.
       I know because 3 tens is less than 6 tens." Go.

Lesson 15:   Explore a situation with more than 9 groups of ten.

205

©2015 Great Minds. eureka-math.org
G2-M3-TE-B2-1.3.1-01.2016

S: 56 is greater than 23. 5 tens are greater than 2 tens.
→ 12 is less than 22 because 1 ten is less than 2 tens.
→ 79 is equal to 79. I know because the tens and ones
are the same.

T: Good. I'm holding our symbols face down. I'll flip one
over, and we'll read it to see which number wins this
round. (Flip over a symbol and show it. This element
of the game encourages students to diversify the
numbers they make.)

T: Who wins?

S: Less than!

T: Yes, the number that is *less than* wins this time.

T: Let's play again. Players, use your digit cards to make
another number.

Continue, following the same sequence.

Concept Development (30 minutes)

Materials: (S) Problem Set

Note: If a document camera is not available for the
Student Debrief, give students poster board for Problem 4.
Students need access to base ten materials (disks, bundles,
and/or blocks) at centers. Do *not* place them at the tables
or explicitly suggest that students use them. This is so that
they learn to use appropriate tools strategically (MP.5).

T: Let's read our 4 problems.

S: (Read.)

T: Partner A, without looking at the paper, retell the
problems to your partner.

T: Partner B, without looking at the paper, retell the
problems, too.

T: Your task in class today is to solve these "pencil
problems" and record your thinking on paper so
that you can share your solution strategies with
another group.

T: Before we begin, does anyone have any
questions?

S: How much time do we have?

**NOTES ON
MULTIPLE MEANS
OF ACTION AND
EXPRESSION:**

Group students carefully to balance
strengths and language support. It
often works well to pair an English
language learner who has excellent
conceptual understanding with a
student who has very good language
but struggles with content. This pairing
tends to foster supportive cooperation.

So that all students participate in
articulating solutions, consider
requiring that the group determine a
different presenter for each problem.
Groups must work together to make
sure that each presenter is prepared to
share the group's work.

EUREKA
MATH™

T: Good question. I will give you time signals. You have 20 minutes in all. I will tell you when you have 15, 10, and 5 minutes left.

T: Make sure to include a statement of your answers. You may begin!

 MP.1 As the students work, circulate. This is their second extended exploration after many days of consecutive teaching. This is a day to stand back and observe them independently making sense of a problem and persevering in solving it (MP.1). Encourage pairs to ask other pairs for help rather than ask the teacher.

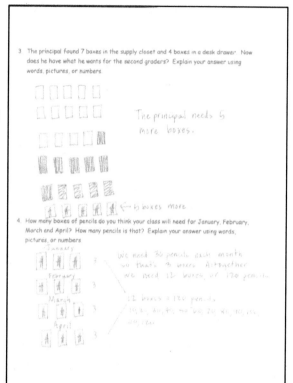

Problem Set (20 minutes)

Students should do their personal best to complete the Problem Set within the allotted 20 minutes. For some classes, it may be appropriate to modify the assignment by specifying which problems they work on first. Some problems do not specify a method for solving. Students should solve these problems using the RDW approach used for Application Problems.

Student Debrief (18 minutes)

Lesson Objective: Explore a situation with more than 9 groups of ten.

The Student Debrief is intended to invite reflection and active processing of the total lesson experience.

Invite students to review their solutions for the Problem Set. They should check work by comparing answers with a partner before going over answers as a class. Look for misconceptions or misunderstandings that can be addressed in the Debrief. Guide students in a conversation to debrief the Problem Set and process the lesson.

T: Bring your work to our Debrief. Partners, find another group with whom to share your work just on Problems 1–3 for now. Explain your solution strategies.

S: (Share with each other.)

T: Let's go over the answers to Problems 1–3. Wait for the signal. Problem 1?

S: 140 pencils.

T: Please give the answer in a full statement, Jeremy.

S: There are 140 pencils in all.

T: Problem 2?

S: The principal needs 160 pencils.

T: What unit are we solving for?

S: Boxes.

Lesson 15: Explore a situation with more than 9 groups of ten.

207

T: So, does 160 pencils answer the question?

S: No.

T: How many boxes does the principal need?

S: 16 boxes.

T: Problem 3. Does the principal have enough pencils?

S: No!

T: How do you know?

S: He found 11 boxes in all. That's 110 pencils. 110 and 140 is less than 300. → He has 140 pencils.
     He found 70 and 40. That's 110. So, put those together, and you have 2 hundreds 5 tens. That's 250.
     Not enough. → You have 2 hundreds, and then 10 and 40 is 50, so it's just 250 and not 300. → He had
     14 boxes. He found 11 boxes. That's 25 boxes, but he needs 5 more to have 30 boxes.

T: Good thinking. He does not have enough pencils. Let's show two different solutions.

T: Now, let's share our work for Problem 4. (Possibly project first the most concrete–pictorial work
     that best supports the mathematical objective.) Tell your partner what you see about how they
     solved the problem.

T: (Allow one or two minutes before continuing.) Now, look at these students' work (show the second
     one down from the top.) Tell your partner what you see about how they solved the problem.

T: (Allow them one or two minutes.) Did both groups get the same answer?

S: No!

T: Talk to your partner about why their answers are different and if both of them can be right.

T: Do you think both of their answers make sense?

S: Yes.

T: Now, think about how each of them solved the problem.

Continue with the analysis of the student work. Get them to observe and analyze similarities and differences.
The last paper should be the most abstract solution. Ask the students to explain the mathematics.

Possibly have them follow up by writing a letter to the principal showing him their ideas and asking his
thinking about the number of pencils to be ordered for their class for the four months. Have them run a sale
or fund drive to make up the difference, assuming the principal was going to order less.

Exit Ticket (3 minutes)

After the Student Debrief, instruct students to complete the Exit Ticket. A review of their work will help with
assessing students' understanding of the concepts that were presented in today's lesson and planning more
effectively for future lessons. The questions may be read aloud to the students.

©2015 Great Minds. eureka-math.org
G2-M3-TE-B2-1.3.1-01.2016

A

Number Correct: _____

Expanded Notation

1.	20 + 1 =	
2.	20 + 2 =	
3.	20 + 3 =	
4.	20 + 9 =	
5.	30 + 9 =	
6.	40 + 9 =	
7.	80 + 9 =	
8.	40 + 4 =	
9.	50 + 5 =	
10.	10 + 7 =	
11.	20 + 5 =	
12.	200 + 30 =	
13.	300 + 40 =	
14.	400 + 50 =	
15.	500 + 60 =	
16.	600 + 70 =	
17.	700 + 80 =	
18.	200 + 30 + 5 =	
19.	300 + 40 + 5 =	
20.	400 + 50 + 6 =	
21.	500 + 60 + 7 =	
22.	600 + 70 + 8 =	

23.	400 + 20 + 5 =	
24.	200 + 60 + 1 =	
25.	200 + 1 =	
26.	300 + 1 =	
27.	400 + 1 =	
28.	500 + 1 =	
29.	700 + 1 =	
30.	300 + 50 + 2 =	
31.	300 + 2 =	
32.	100 + 10 + 7 =	
33.	100 + 7 =	
34.	700 + 10 + 5 =	
35.	700 + 5 =	
36.	300 + 40 + 7 =	
37.	300 + 7 =	
38.	500 + 30 + 2 =	
39.	500 + 2 =	
40.	2 + 500 =	
41.	2 + 600 =	
42.	2 + 40 + 600 =	
43.	3+ 10 + 700 =	
44.	8 + 30 + 700 =	

Lesson 15:   Explore a situation with more than 9 groups of ten.

209

B

Number Correct: _____

Improvement: _____

Expanded Notation

1.	10 + 1 =	
2.	10 + 2 =	
3.	10 + 3 =	
4.	10 + 9 =	
5.	20 + 9 =	
6.	30 + 9 =	
7.	70 + 9 =	
8.	30 + 3 =	
9.	40 + 4 =	
10.	80 + 7 =	
11.	90 + 5 =	
12.	100 + 20 =	
13.	200 + 30 =	
14.	300 + 40 =	
15.	400 + 50 =	
16.	500 + 60 =	
17.	600 + 70 =	
18.	300 + 40 + 5 =	
19.	400 + 50 + 6 =	
20.	500 + 60 + 7 =	
21.	600 + 70 + 8 =	
22.	700 + 80 + 9 =	

23.	500 + 30 + 6 =	
24.	300 + 70 + 1 =	
25.	300 + 1 =	
26.	400 + 1 =	
27.	500 + 1 =	
28.	600 + 1 =	
29.	900 + 1 =	
30.	400 + 60 + 3 =	
31.	400 + 3 =	
32.	100 + 10 + 5 =	
33.	100 + 5 =	
34.	800 + 10 + 5 =	
35.	800 + 5 =	
36.	200 + 30 + 7 =	
37.	200 + 7 =	
38.	600 + 40 + 2 =	
39.	600 + 2 =	
40.	2 + 600 =	
41.	3 + 600 =	
42.	3 + 40 + 600 =	
43.	5 + 10 + 800 =	
44.	9 + 20 + 700 =	

Lesson 15:   Explore a situation with more than 9 groups of ten.

EUREKA
MATH™

©2015 Great Minds. eureka-math.org
G2-M3-TE-B2-1.3.1-01.2016

Names _____ and _____ Date _____

Pencils come in boxes of 10.

There are 14 boxes.

1. How many pencils are there in all? Explain your answer using words, pictures, or numbers.

2. The principal wants to have 300 pencils for the second graders for October, November, and December. How many more boxes of pencils does he need? Explain your answer using words, pictures, or numbers.

Lesson 15: Explore a situation with more than 9 groups of ten.

211

©2015 Great Minds. eureka-math.org
G2-M3-TE-B2-1.3.1-01.2016

3. The principal found 7 boxes in the supply closet and 4 boxes in a desk drawer. Now does he have what he wants for the second graders? Explain your answer using words, pictures, or numbers.

4. How many boxes of pencils do you think your class will need for January, February, March, and April? How many pencils is that? Explain your answer using words, pictures, or numbers.

EUREKA
MATH

Name _____   Date _____

Think about the different strategies and tools your classmates used to answer the pencil question. Explain a strategy you liked that is <u>different</u> from yours using words, pictures, or numbers.

Lesson 15:   Explore a situation with more than 9 groups of ten.

213

©2015 Great Minds. eureka-math.org
G2-M3-TE-B2-1.3.1-01.2016

Name _____ Date _____

Pencils come in boxes of 10.

1. How many boxes should Erika buy if she needs 127 pencils?

2. How many pencils will Erika have left over after she gets what she needs out of the boxes?

3. How many more pencils does she need to have 200 pencils?

EUREKA
MATH

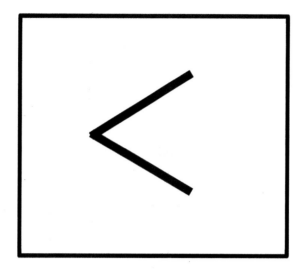

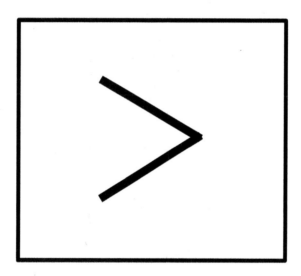

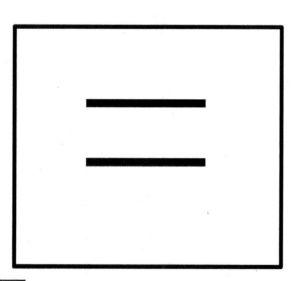

<, >, = symbol cards

Lesson 15:   Explore a situation with more than 9 groups of ten.

215

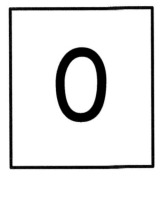

0	1	2
3	4	5
6	7	8

9

_____

digit cards 0–9

Lesson 15:   Explore a situation with more than 9 groups of ten.

EUREKA
MATH

Mathematics Curriculum

2
GRADE

Topic F

Comparing Two Three-Digit Numbers

2.NBT.4

Focus Standard:	2.NBT.4	Compare two three-digit numbers based on meanings of the hundreds, tens, and ones digits, using >, =, and < symbols to record the results of comparisons.
Instructional Days:	3	
Coherence -Links from:	G1–M6	Place Value, Comparison, Addition and Subtraction to 100
-Links to:	G2–M4	Addition and Subtraction Within 200 with Word Problems to 100

Place value disks make comparison of numbers very easy. *More than* and *less than* lead to addition and subtraction in the next module. In Lesson 16, students compare numbers using the symbols <, >, and = on the place value chart. Next, students advance to comparing different forms (**2.NBT.4**), and finally, in Lesson 18, they apply their comparison and place value skills to order more than two numbers in different forms.

A Teaching Sequence Toward Mastery of Comparing Two Three-Digit Numbers
Objective 1: Compare two three-digit numbers with <, >, and =.   (Lesson 16)
Objective 2: Compare two three-digit numbers with <, >, and = when there are more than 9 ones or 9 tens.   (Lesson 17)
Objective 3: Order numbers in different forms. (Optional)   (Lesson 18)

Lesson 16

Objective: Compare two three-digit numbers using <, >, and =.

Suggested Lesson Structure

■ Fluency Practice   (12 minutes)
▨ Application Problem   (8 minutes)
☐ Concept Development   (30 minutes)
■ Student Debrief   (10 minutes)

   Total Time   **(60 minutes)**

Fluency Practice (12 minutes)

▪ Sprint: Sums—Crossing Ten **2.OA.2**   (12 minutes)

Sprint: Sums—Crossing Ten (12 minutes)

Materials: (S) Sums—Crossing Ten Sprint

In Topics F and G for the next 6 days of instruction, a blitz is done on addition and subtraction sums in preparation for Module 4. As the beginning of Module 4 draws near, the goal is to energize and hone students' addition and subtraction facts before getting there.

Application Problem (8 minutes)

At recess Diane skipped rope 65 times without stopping. Peter skipped rope 20 times without stopping. How many more times did Diane skip rope than Peter?

Note: Lead students as necessary through the sequence of questions they need to internalize:

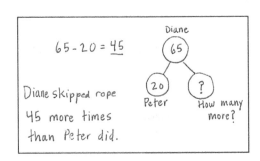

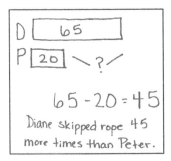

- ▪ What do you see?
- ▪ Can you draw something?
- ▪ What can you draw?
- ▪ What conclusions can you make from your drawing?
- T: Use your RDW process. (Allow time to work.)
- T: I notice some of you used addition, and some of you used subtraction to find the answer.

EUREKA
MATH™

T: Who would like to share what they wrote?

S: $65 - 20 =$ ___ $\rightarrow 20 +$ ___ $= 65$.

T: Were you missing a part or the whole?

S: A part.

T: Turn and talk to your partner about what is the missing part in the story of Diane and Peter.

S: It's the number of jumps Diane did that Peter didn't do. $\rightarrow$ It's how many more Peter had to do to have the same number of jumps as Diane.

T: We are comparing. What did you learn in Grade 1 about comparing and subtraction?

S: We learned that to compare, you subtract, because you're finding the part that is missing.

T: Excellent. Let's look at that missing part in two excellent drawings made by your friends. See if you can find them. Talk to your partner.

Let the students point to the missing part in the drawings and really make that connection between the number sentence and the missing part.

Concept Development (30 minutes)

Materials: (S) Unlabeled hundreds place value chart (Lesson 8 Template), place value disks (2 hundreds, 7 tens, and 7 ones), personal white board, number comparison (Template)

Concrete (6 minutes)

T: Slide the place value chart inside your personal white boards.

T: Use place value disks to show 74 on your place value chart.

S: (Show.)

T: Which disks did you use from greatest to least?

S: Tens and ones.

T: Add 1 disk so the number becomes 174.

T: (Show.) What did you add?

S: A hundred.

T: Which number is greater? 74 or 174?

S: 174.

T: Let's state that as a sentence.

S: 174 is greater than 74.

T: Change your disks to show 105.

T: (Show.) Which disks did you use from greatest to least?

S: Hundreds and ones.

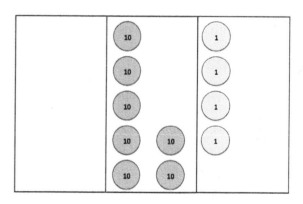

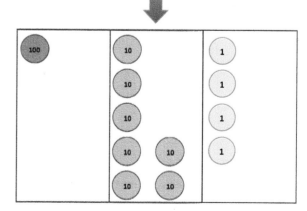

T: Now, make the number 135.

T: (Show.) Which number is less, 105 or 135?

S: 105.

T: Say it as a sentence.

S: 105 is less than 135.

T: Talk to your partner. How can you tell?
(Circulate and listen.)

T: Let's hear some of your good ideas.
(Choose students to share based on their thinking.)

S: 105 has fewer tens. → 135 has 3 tens. 105 has no
tens. → 105 has 10 tens. 135 has 13 tens.

T: Show 257 on your place value chart.

T: (Show.) Change it to show 250.

T: (Show.) Which number is greater, 257 or 250?

S: 257.

T: Say it as a sentence.

S: 257 is greater than 250.

T: How do you know?

S: We took away the ones, and 257 got smaller.

Continue with other examples until students gain proficiency.

Pictorial (12 minutes)

Materials: (S) Number comparison pre-cut (template)

T: Take two minutes to draw each number using hundreds, tens, and ones.

T: Compare with your partner. How are your drawings alike?

S: (Compare.)

T: Look carefully at our three numbers. Which is greatest?

472   274   724

S: 724.

T: Turn and tell your partner how you know. (Encourage precise explanation.)

NOTES ON
MULTIPLE MEANS
OF ACTION AND
EXPRESSION:

Support students by creating or posting
a chart of words. It might be as simple
as:

smaller   bigger
smallest   biggest
less than   greater than
least   greatest
<   >

This lesson introduces the sentence
frame:

_____ strategy is _____.

Students benefit from articulating how
another student thought about or
solved a problem. Listening is essential
to learning a second language. When
students hear the familiar names of
their peers, they sense a classroom
community that is personal, respectful,
and caring. This positive feeling hooks
them into the lesson.

220   Lesson 16:   Compare two three-digit numbers using <, >, and =.

EUREKA
MATH

©2015 Great Minds. eureka-math.org
G2-M3-TE-B2-1.3.1-01.2016

T: Some students compared the number of tens, and others compared the number of hundreds.

T: Turn and tell your partner why comparing units might help us figure out which number is greatest.

While circulating, identify exemplary explanations.

S: It works because there are the most hundreds in 724. → Hundreds come first, so it's easiest to compare them first. → There are more tens inside the hundreds. 724 really has 72 tens, and 274 has only 27.

T: Quite a few of you have excellent explanations. Melanie, will you share your thinking?

S: Hundreds are the biggest unit. So, if a number has 7 hundreds and the other has only 4, you already know that the one with 7 has to be greatest.

T: If we use Melanie's strategy, which number is least?

S: 274.

T: Anthony, will you share how you compare tens? After he shares, I'll ask everyone to retell his idea.

S: 274 has more tens in the tens place than 724, but the number is not greater. I said you have to remember to think about *all* the tens. 724 really has 72 tens, and 274 really has 27 tens.

T: (Write the sentence frame, "Anthony's strategy is _____.") Use the frame to retell Anthony's strategy to your partner.

S: (Retell.)

T: Use Anthony's strategy. Name just the tens, and say the three numbers from greatest to least.

S: 72 tens, 47 tens, 27 tens.

T: Good. Use the symbols < or > to write a number sentence with all three numbers at the bottom of Problem Set 1.

S: (Write.)

T: Check your partner's work. It might look different from yours, but make sure you agree it's true.

T: Look at 341 and 329 (write these numbers on the board). The number of hundreds is the same. What would you do to compare then?

S: Look at the tens. 4 tens is more than 2 tens.

©2015 Great Minds. eureka-math.org
G2-M3-TE-B2-1.3.1-01.2016

Problem Set (12 minutes)

Materials: (S) Problem Set

Students should do their personal best to complete the Problem Set within the allotted 12 minutes. For some classes, it may be appropriate to modify the assignment by specifying which problems they work on first. Some problems do not specify a method for solving. Students should solve these problems using the RDW approach used for Application Problems.

Instruct students to complete the Problem Set by drawing values on the place value chart as specified and answering the included questions.

Student Debrief (10 minutes)

Lesson Objective: Compare two three-digit numbers using <, >, and =.

Materials: (S) Problem Set

The Student Debrief is intended to invite reflection and active processing of the total lesson experience.

Invite students to review their solutions for the Problem Set. They should check work by comparing answers with a partner before going over answers as a class. Look for misconceptions or misunderstandings that can be addressed in the Debrief. Guide students in a conversation to debrief the Problem Set and process the lesson.

- T: Bring Problem Set to our Debrief.
- T: Check your work carefully with a partner as I circulate.
- T: (Allow two minutes.) Which problems were hard for you?
- S: (Respond.)
- T: Look at the threes in 132 and 312.
       What is the difference between them?
- S: One is in the tens place, and the other is in the hundreds place.

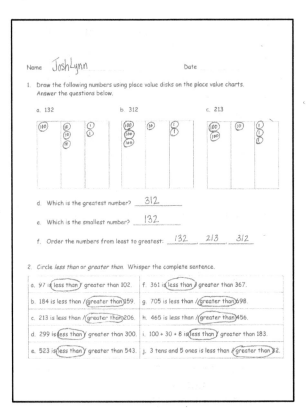

Lesson 16:   Compare two three-digit numbers using <, >, and =.

EUREKA MATH™

T: You noticed place value. How did place value help you to compare the numbers on the Problem Set in Problem 3 (e)?

S: 908 and 980 sound almost the same, but if you notice what place the 8 is in, you know that 8 tens is more than 8 ones.

T: Some problems like 3(e) used word form or unit form. Could you still use place value to compare? How did you do it?

S: I just wrote the numbers in standard form. Then, it was easy to look at them and see the numbers in their places.

T: Look back at each section of our Problem Set. What was the same about your task in each one?

S: We always had to compare!

T: Now, think about your strategy for comparing. Turn and tell your strategy to your partner. Say, "My strategy is…"

S: My strategy is to compare numbers by looking at hundreds, tens, and ones. → My strategy is to compare places. → My strategy is to compare numbers using place value.

T: Write your strategy on your Problem Set so you're sure to remember it. (Allow time to write.)

T: Share with your partner about Noah and Charlie's problem and your thinking about who is correct.

S: (Share answers to Problem 4.)

T: What materials in our classroom could we use to prove who is correct?

S: The bundles of sticks. → The blocks. → The dollar bills. → Place value disks.

T: True. When we see materials, sometimes it makes the comparison so obvious!

Exit Ticket (3 minutes)

After the Student Debrief, instruct students to complete the Exit Ticket. A review of their work will help with assessing students' understanding of the concepts that were presented in today's lesson and planning more effectively for future lessons. The questions may be read aloud to the students.

©2015 Great Minds. eureka-math.org
G2-M3-TE-B2-1.3.1-01.2016

A

Number Correct: _____

Sums—Crossing Ten

1.	9 + 1 =	
2.	9 + 2 =	
3.	9 + 3 =	
4.	9 + 9 =	
5.	8 + 2 =	
6.	8 + 3 =	
7.	8 + 4 =	
8.	8 + 9 =	
9.	9 + 1 =	
10.	9 + 4 =	
11.	9 + 5 =	
12.	9 + 8 =	
13.	8 + 2 =	
14.	8 + 5 =	
15.	8 + 6 =	
16.	8 + 8 =	
17.	9 + 1 =	
18.	9 + 7 =	
19.	8 + 2 =	
20.	8 + 7 =	
21.	9 + 1 =	
22.	9 + 6 =	

23.	7 + 3 =	
24.	7 + 4 =	
25.	7 + 5 =	
26.	7 + 9 =	
27.	6 + 4 =	
28.	6 + 5 =	
29.	6 + 6 =	
30.	6 + 9 =	
31.	5 + 5 =	
32.	5 + 6 =	
33.	5 + 7 =	
34.	5 + 9 =	
35.	4 + 6 =	
36.	4 + 7 =	
37.	4 + 9 =	
38.	3 + 7 =	
39.	3 + 9 =	
40.	5 + 8 =	
41.	2 + 8 =	
42.	4 + 8 =	
43.	1 + 9 =	
44.	2 + 9 =	

Lesson 16:   Compare two three-digit numbers using <, >, and =.

EUREKA
MATH™

B

Sums—Crossing Ten

Number Correct: _____

Improvement: _____

1.	8 + 2 =			23.	7 + 3 =	
2.	8 + 3 =			24.	7 + 4 =	
3.	8 + 4 =			25.	7 + 5 =	
4.	8 + 8 =			26.	7 + 8 =	
5.	9 + 1 =			27.	6 + 4 =	
6.	9 + 2 =			28.	6 + 5 =	
7.	9 + 3 =			29.	6 + 6 =	
8.	9 + 8 =			30.	6 + 8 =	
9.	8 + 2 =			31.	5 + 5 =	
10.	8 + 5 =			32.	5 + 6 =	
11.	8 + 6 =			33.	5 + 7 =	
12.	8 + 9 =			34.	5 + 8 =	
13.	9 + 1 =			35.	4 + 6 =	
14.	9 + 4 =			36.	4 + 7 =	
15.	9 + 5 =			37.	4 + 8 =	
16.	9 + 9 =			38.	3 + 7 =	
17.	9 + 1 =			39.	3 + 9 =	
18.	9 + 7 =			40.	5 + 9 =	
19.	8 + 2 =			41.	2 + 8 =	
20.	8 + 7 =			42.	4 + 9 =	
21.	9 + 1 =			43.	1 + 9 =	
22.	9 + 6 =			44.	2 + 9 =	

Lesson 16:   Compare two three-digit numbers using <, >, and =.

225

Name _____ Date _____

1. Draw the following numbers using place value disks on the place value charts. Answer the questions below.

   a. 132

   b. 312

   c. 213

   d. Which is the greatest number? _____

   e. Which is the least number? _____

   f. Order the numbers from least to greatest: _____, _____, _____

2. Circle *less than* or *greater than*. Whisper the complete sentence.

a. 97 is less than / greater than 102.	f. 361 is less than / greater than 367.
b. 184 is less than / greater than 159.	g. 705 is less than / greater than 698.
c. 213 is less than / greater than 206.	h. 465 is less than / greater than 456.
d. 299 is less than / greater than 300.	i. 100 + 30 + 8 is less than / greater than 183.
e. 523 is less than / greater than 543.	j. 3 tens and 5 ones is less than / greater than 32.

EUREKA MATH™

3. Write >, <, or =. Whisper the complete number sentences as you work.

a. 900 $\bigcirc$ 899

b. 267 $\bigcirc$ 269

c. 537 $\bigcirc$ 527

d. 419 $\bigcirc$ 491

e. 908 $\bigcirc$ nine hundred eighty

f. 130 $\bigcirc$ 80 + 40

g. Two hundred seventy-one $\bigcirc$ 70 + 200 + 1

h. 500 + 40 $\bigcirc$ 504

i. 10 tens $\bigcirc$ 101

j. 4 tens 2 ones $\bigcirc$ 30 + 12

k. 36 – 10 $\bigcirc$ 2 tens 5 ones

4. Noah and Charlie have a problem.

Noah thinks 42 tens is <u>less than</u> 390.

Charlie thinks 42 tens is <u>greater than</u> 390.

Who is correct? Explain your thinking below.

_____

_____

_____

©2015 Great Minds. eureka-math.org
G2-M3-TE-B2-1.3.1-01.2016

Name _____   Date _____

Write >, <, or =.

1. 499 ◯ 500

2. 179 ◯ 177

3. 431 ◯ 421

4. 703 ◯ seven hundred three

5. 2 hundred 70 ones ◯ 70 + 200 + 1

6. 300 + 60 ◯ 306

7. 4 tens 2 ones ◯ 30 + 12

8. 3 tens 7 ones ◯ 45 – 10

Lesson 16:   Compare two three-digit numbers using <, >, and =.

EUREKA
MATH™

Name _____   Date _____

1. Draw the following numbers using place value disks on the place value charts. Answer the questions below.

a. 241

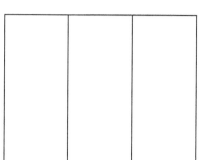

b. 412

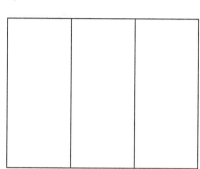

c. 124

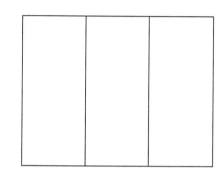

d. Order the numbers from least to greatest: _____, _____, _____

2. Circle *less than* or *greater than*. Whisper the complete sentence.

a. 112 is less than / greater than 135.	d. 475 is less than / greater than 457.
b. 152 is less than / greater than 157.	e. 300 + 60 + 5 is less than / greater than 635.
c. 214 is less than / greater than 204.	f. 4 tens and 2 ones is less than / greater than 24.

3. Write >, <, or =.

a. 100 ◯ 99   e. 150 ◯ 90 + 50

b. 316 ◯ 361   f. 9 tens 6 ones ◯ 92

c. 523 ◯ 525   g. 6 tens 8 ones ◯ 50 + 18

d. 602 ◯ six hundred two   h. 84 – 10 ◯ 7 tens 5 ones

Name _____ Date _____

472

274

724

Name _____ Date _____

472

274

724

Name _____ Date _____

472

274

724

number comparison template

EUREKA MATH

Lesson 17

Objective: Compare two three-digit numbers using <, >, and = when there are more than 9 ones or 9 tens.

Suggested Lesson Structure

■ Fluency Practice   (12 minutes)
▨ Application Problem   (8 minutes)
▢ Concept Development   (30 minutes)
■ Student Debrief   (10 minutes)

Total Time   **(60 minutes)**

Fluency Practice (12 minutes)

▪ Sprint: Sums—Crossing Ten **2.OA.2**   (12 minutes)

Sprint: Sums—Crossing Ten (12 minutes)

Materials: (S) Sums—Crossing Ten Sprint

Day 2 of our Sums and Differences blitz continues with another Sprint on sums and differences to 20.

  T: (After students have taken the Sprint.) Tomorrow, we are going to do the exact same sprint. If you wish to take this home and study or practice to see if you can do the problems more skillfully, do so!

  T: Take a moment to analyze the Sprint with your partner. It is arranged from the easiest problems to the hardest.

  S: It starts with the 9+ facts. That's easy! You make a ten! → Or, I just do it like a 10+ and do
1 less. → Yeah, and then it goes to the 10+ facts. Those are super easy!

  T: Raise your hand if you think you might do better tomorrow.

NOTES ON
MULTIPLE MEANS
OF ENGAGEMENT:

The Sprint can be highly motivating for students working below grade level if they can stop comparing their performance to others and really take note of personal improvement.

Privately record the student's score each day. Ask her if she practiced. Celebrate improvement, even if by one question, the moment it occurs. Do this discreetly until the student is confident that she is capable of consistent success.

For English language learners, this is a real chance to shine. Math is a universal language, so calculations offer no impediment. Let them savor the adrenaline of academic success.

Lesson 17:   Compare two three-digit numbers using <, >, and = when there are
                more than 9 ones or 9 tens.

231

©2015 Great Minds. eureka-math.org
G2-M3-TE-B2-1.3.1-01.2016

Application Problem (8 minutes)

Walking on the beach on Tuesday, Darcy collected 35 rocks. The day before, she collected 28. How many fewer rocks did she collect on Monday than on Tuesday?

T: Use your RDW process. What do you see?

S: Her 35 rocks from Tuesday and her 28 rocks from Monday.

T: Can you draw something?

S: Yes!

T: What can you draw?

S: The rocks!

T: I'm only going to give you two minutes to draw. Can you think of efficient shortcuts so that you don't have to draw all the rocks?

S: Yes!

T: Okay. Do so.

S: (Show.)

T: What problem did you write to find the answer?

S: 35 − 28 = ___ → 28 + ___ = 35.

T: Darcy is comparing. Talk to your partner about what she is comparing.

T $\boxed{ 35 }$

M $\boxed{ 28 }$?

$35 - 28 = \underline{7}$

$28 + 2 = 30$
$30 + 5 = 35$

Darcy collected 7 fewer rocks on Monday.

Lead the students in a conversation about subtraction and comparison. Yes, they are finding a missing part. As time permits, look at different examples of student work.

Concept Development (30 minutes)

Materials: (S) Unlabeled hundreds place value chart (Lesson 8 Template), place value disks (9 hundreds, tens, and ones) per student; one set of pre-cut <, >, = symbol cards (Lesson 15 Template 1) per pair

Concrete (5 minutes)

Note: Have the student on the left be Partner A.

T: Partner A, show 124 on your place value chart. Partner B, show 824.

S: (Show.)

T: Compare numbers. Place a symbol from the set between your charts to make a true statement. Read the statement.

S: (Place <.) 124 is less than 824.

NOTES ON MULTIPLE MEANS OF REPRESENTATION:

While students are familiar with the language of tens and ones, they may feel overwhelmed when asked to manipulate two units at once. Support English language learners by writing the mathematical equivalent to the words on the board.

Partner A	Partner B
− 4 tens 4 ones	+ 2 tens 6 ones
5 tens 6 ones	15 tens 6 ones
+ 7 tens 5 ones	− 2 tens 5 ones

Point to the symbols. As students manipulate the place value disks, the visual, kinesthetic, and auditory are coming together powerfully.

232   Lesson 17:   Compare two three-digit numbers using <, >, and = when there are more than 9 ones or 9 tens.

EUREKA MATH

©2015 Great Minds. eureka-math.org
G2-M3-TE-B2-1.3.1-01.2016

T: Partner A, add 7 tens to your number. Partner B, take 7 hundreds from your number.

S: (Show.)

T: Compare. Choose the symbol to go between your charts. Read the statement.

S: (Place >.) 194 is greater than 124.

T: Partner A, take 4 tens 4 ones from your number. Partner B, add 2 tens 6 ones to yours.

T: Compare numbers. Choose the symbol. Read the statement.

S: (Place =.) 150 equals 150.

T: How many tens in 150?

S: 15.

T: Partner A, show 5 tens 6 ones. Partner B, show 15 tens 6 ones.

S: (Show.)

T: Compare numbers, and place your symbol. Read the statement, naming just tens and ones.

S: (Place <.) 5 tens 6 ones is less than 15 tens 6 ones.

T: Partner A, add 7 tens 5 ones to your number. Partner B, take 2 tens 5 ones from your number.

S: (Show.)

T: Compare numbers, and place your symbol. Read the statement naming just tens and ones.

S: (Place =.) 13 tens 1 one equals 13 tens 1 one.

T: (Write 113 on the board.) Read my number in standard form.

S: 113.

T: Is my number greater than, less than, or equal to yours? Decide with your partner, then hold up a symbol.

S: (Hold up <.)

T: Say the number sentence. Say my number in standard form, and name yours with tens and ones.

S: 113 is less than 13 tens 1 one.

Pictorial (10 minutes)

Materials: (T) 2 unlabeled hundreds place value charts (Lesson 8 Template) for projection, place value disks (17 hundreds, 15 tens, 15 ones) (S) Personal white board

As an alternative to projecting the place value charts, the teacher may slip place value chart templates into a personal white board and use a marker to draw.

T: (Show 55 on the first chart.) Write this number in standard form. Turn your board horizontally so you have room to write a second number beside it.

S: (Write 55.)

Lesson 17:   Compare two three-digit numbers using <, >, and = when there are
            more than 9 ones or 9 tens.

233

©2015 Great Minds. eureka-math.org
G2-M3-TE-B2-1.3.1-01.2016

T: (Show 50 on the second chart.) Now, write this number in unit form.

S: (Write 5 tens.)

T: Draw a symbol comparing the numbers. Read the number sentence.

S: (Draw >.) 55 is greater than 5 tens.

T: Good. Erase. (Show 273 on the first chart.) Write in unit form, naming only tens and ones.

S: (Write 27 tens 3 ones.)

T: (Show 203 on the second chart.) Write in expanded form.

S: (Write 200 + 3 or 3 + 200.)

T: Draw a symbol to compare the numbers, and then read the number sentence.

MP.6

S: (Draw >.) 27 tens 3 ones is greater than 200 + 3.

T: Nice. Erase. (Show 406 on the first chart.) Write in word form.

S: (Write four hundred six.)

T: (Show 436 on the second chart.) Write in expanded form.

S: (Write 400 + 30 + 6 or a variation on that order.)

T: Draw a symbol and read.

S: (Draw <.) Four hundred six is less than 400 + 30 + 6.

T: (Show 920 on the first chart.) Write in standard form.

S: (Show 920.)

T: (Show 880 on the second chart.) Write in unit form, naming only tens and ones.

S: (Write 88 tens.)

T: Draw a symbol and read.

S: (Draw >.) 920 is greater than 88 tens.

T: Good. On your board, write *+ 4 tens* after 88 tens. Solve. Change the symbol if you need to.

S: (Work.)

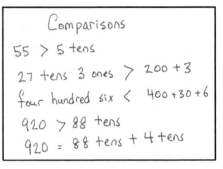

Comparisons
55 > 5 tens
27 tens 3 ones > 200 + 3
four hundred six < 400 + 30 + 6
920 > 88 tens
920 = 88 tens + 4 tens

T: Partner A, show your partner how you solved 88 tens + 4 tens.

S: I looked at the teacher's picture. I started with 880 and counted by tens 4 times—890, 900, 910, 920. → Oops, I changed it to 884. → I did 88 + 4. Then, I got 92, so I knew it changed to 92 tens.

T: Partner B, talk to your partner about what happened to the symbol. Read the number sentence.

S: Once they were both 92 tens I changed the symbol to =. Now, it says 92 tens equals 92 tens.

©2015 Great Minds. eureka-math.org
G2-M3-TE-B2-1.3.1-01.2016

EUREKA
MATH™

Problem Set (15 minutes)

Students should do their personal best to complete the Problem Set within the allotted 15 minutes. For some classes, it may be appropriate to modify the assignment by specifying which problems they work on first. Some problems do not specify a method for solving. Students should solve these problems using the RDW approach used for Application Problems.

Review the Problem Set instructions with students.

Student Debrief (10 minutes)

Lesson Objective: Compare two three-digit numbers using <, >, and = when there are more than 9 ones or 9 tens.

The Student Debrief is intended to invite reflection and active processing of the total lesson experience.

Invite students to review their solutions for the Problem Set. They should check work by comparing answers with a partner before going over answers as a class. Look for misconceptions or misunderstandings that can be addressed in the Debrief. Guide students in a conversation to debrief the Problem Set and process the lesson.

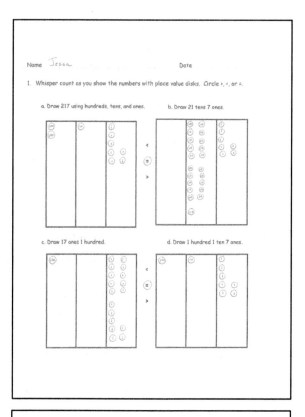

- T: Bring your Problem Set to our Debrief.
- S: Check your work carefully with a partner as I circulate. Put a little star next to the ones that were hard.
- T: (Allow two minutes.) Which ones were hard for you?
- S: Problem 2(h) was hard!
- T: Tell us what made it difficult.
- S: I thought doing 47 tens + 23 tens was tricky because it's a lot of tens to draw.
- T: That's true! Drawing takes a while. Can someone share a more efficient strategy?
- S: I used the 3 from 23 to make a ten with 47. That was 50. Then, it was just 50 + 20. Easy. 70 tens!
- T: Turn and talk to your partner about Hyun-Mee's strategy for quickly solving 47 tens + 23 tens.
- S: She made a ten! I guess you could just do 7 + 3 to get a ten too and then add 4 tens, 2 tens, and 1 ten.

Lesson 17: Compare two three-digit numbers using <, >, and = when there are more than 9 ones or 9 tens.

©2015 Great Minds. eureka-math.org
G2-M3-TE-B2-1.3.1-01.2016

235

T: What's another question you starred?

S: Problem 3(g). I didn't notice the units are mixed up in the number that's unit form. I thought it was 964 instead of 649.

T: What will you do differently to avoid that mistake next time?

S: I need to slow down and read more carefully. I wasn't really paying attention to units, just to order.

T: Thanks for pointing that out, Austin. Thumbs up if you made that mistake on one of the problems.

S: (Several students show thumbs up.)

T: Did anyone have a strategy for paying attention to units?

S: As I read the problems, I just wrote the numbers in standard form. That way I didn't get messed up.

T: Nice. It's important to have little strategies for helping yourself.

T: Head back to your seats to complete your Exit Ticket.

Exit Ticket (3 minutes)

After the Student Debrief, instruct students to complete the Exit Ticket. A review of their work will help with assessing students' understanding of the concepts that were presented in today's lesson and planning more effectively for future lessons. The questions may be read aloud to the students.

Lesson 17:   Compare two three-digit numbers using <, >, and = when there are more than 9 ones or 9 tens.

EUREKA MATH

A

Number Correct: _____

Sums—Crossing Ten

1.	9 + 2 =		23.	4 + 7 =	
2.	9 + 3 =		24.	4 + 8 =	
3.	9 + 4 =		25.	5 + 6 =	
4.	9 + 7 =		26.	5 + 7 =	
5.	7 + 9 =		27.	3 + 8 =	
6.	10 + 1 =		28.	3 + 9 =	
7.	10 + 2 =		29.	2 + 9 =	
8.	10 + 3 =		30.	5 + 10 =	
9.	10 + 8 =		31.	5 + 8 =	
10.	8 + 10 =		32.	9 + 6 =	
11.	8 + 3 =		33.	6 + 9 =	
12.	8 + 4 =		34.	7 + 6 =	
13.	8 + 5 =		35.	6 + 7 =	
14.	8 + 9 =		36.	8 + 6 =	
15.	9 + 8 =		37.	6 + 8 =	
16.	7 + 4 =		38.	8 + 7 =	
17.	10 + 5 =		39.	7 + 8 =	
18.	6 + 5 =		40.	6 + 6 =	
19.	7 + 5 =		41.	7 + 7 =	
20.	9 + 5 =		42.	8 + 8 =	
21.	5 + 9 =		43.	9 + 9 =	
22.	10 + 6 =		44.	4 + 9 =	

Lesson 17:   Compare two three-digit numbers using <, >, and = when there are
more than 9 ones or 9 tens.

237

©2015 Great Minds. eureka-math.org
G2-M3-TE-B2-1.3.1-01.2016

B

Number Correct: _____

Improvement: _____

Sums—Crossing Ten

1.	10 + 1 =	
2.	10 + 2 =	
3.	10 + 3 =	
4.	10 + 9 =	
5.	9 + 10 =	
6.	9 + 2 =	
7.	9 + 3 =	
8.	9 + 4 =	
9.	9 + 8 =	
10.	8 + 9 =	
11.	8 + 3 =	
12.	8 + 4 =	
13.	8 + 5 =	
14.	8 + 7 =	
15.	7 + 8 =	
16.	7 + 4 =	
17.	10 + 4 =	
18.	6 + 5 =	
19.	7 + 5 =	
20.	9 + 5 =	
21.	5 + 9 =	
22.	10 + 8 =	

23.	5 + 6 =	
24.	5 + 7 =	
25.	4 + 7 =	
26.	4 + 8 =	
27.	4 + 10 =	
28.	3 + 8 =	
29.	3 + 9 =	
30.	2 + 9 =	
31.	5 + 8 =	
32.	7 + 6 =	
33.	6 + 7 =	
34.	8 + 6 =	
35.	6 + 8 =	
36.	9 + 6 =	
37.	6 + 9 =	
38.	9 + 7 =	
39.	7 + 9 =	
40.	6 + 6 =	
41.	7 + 7 =	
42.	8 + 8 =	
43.	9 + 9 =	
44.	4 + 9 =	

Lesson 17: Compare two three-digit numbers using <, >, and = when there are more than 9 ones or 9 tens.

EUREKA MATH™

©2015 Great Minds. eureka-math.org
G2-M3-TE-B2-1.3.1-01.2016

Name _____   Date _____

1. Whisper count as you show the numbers with place value disks. Circle >, <, or =.

    a. Draw 217 using hundreds, tens, and ones.   b. Draw 21 tens and 7 ones.

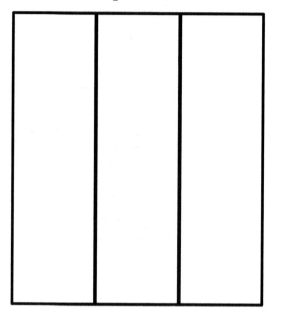

<
=
>

    c. Draw 1 hundred and 17 ones.   d. Draw 1 hundred 1 ten and 7 ones.

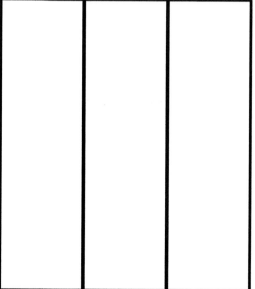

<
=
>

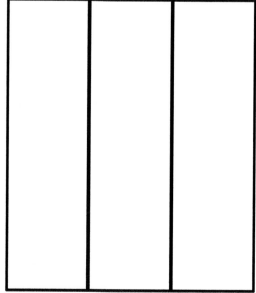

EUREKA
MATH™

Lesson 17:   Compare two three-digit numbers using <, >, and = when there are
                more than 9 ones or 9 tens.

239

2. Circle less than (<), equal to (=), or greater than (>). Whisper the complete sentence.

a. 9 tens is _____ 88.

```
┌──────────────────┐
│   less than       │
│   equal to        │
│   greater than    │
└──────────────────┘
```

b. 132 is _____ 13 tens 2 ones.

```
┌──────────────────┐
│   less than       │
│   equal to        │
│   greater than    │
└──────────────────┘
```

c. 102 is _____ 15 tens 2 ones.

```
┌──────────────────┐
│   less than       │
│   equal to        │
│   greater than    │
└──────────────────┘
```

d. 199 is _____ 20 tens

```
┌──────────────────┐
│   less than       │
│   equal to        │
│   greater than    │
└──────────────────┘
```

e. 62 tens 3 ones is | <   =   > | 623.

f. 80 + 700 + 2 is | <   =   > | eight hundred seventy-two.

g. 8 + 600 is | <   =   > | 68 tens.

h. Seven hundred thirteen is | <   =   > | 47 tens + 23 tens.

i. 18 tens + 4 tens is | <   =   > | 29 tens – 5 tens.

j. 300 + 40 + 9 is | <   =   > | 34 tens.

Compare two three-digit numbers using <, >, and = when there are more than 9 ones or 9 tens.

EUREKA
MATH™

3. Write >, <, or =.

   a. 99 ◯ 10 tens

   b. 116 ◯ 11 tens 5 ones

   c. 2 hundreds 37 ones ◯ 237

   d. Three hundred twenty ◯ 34 tens

   e. 5 hundreds 2 tens 4 ones ◯ 53 tens

   f. 104 ◯ 1 hundred 4 tens

   g. 40 + 9 + 600 ◯ 9 ones 64 tens

   h. 700 + 4 ◯ 74 tens

   i. Twenty-two tens ◯ Two hundreds twelve ones

   j. 7 + 400 + 20 ◯ 42 tens 7 ones

   k. 5 hundreds 24 ones ◯ 400 + 2 + 50

   l. 69 tens + 2 tens ◯ 710

   m. 20 tens ◯ two hundred ten ones

   n. 72 tens – 12 tens ◯ 60

   o. 84 tens + 10 tens ◯ 9 hundreds 4 ones

   p. 3 hundreds 21 ones ◯ 18 tens + 14 tens

Lesson 17:   Compare two three-digit numbers using <, >, and = when there are
             more than 9 ones or 9 tens.

241

©2015 Great Minds. eureka-math.org
G2-M3-TE-B2-1.3.1-01.2016

Name _____   Date _____

1. Whisper count as you show the numbers with place value disks. Circle >, <, or =.

    a. Draw 142 using hundreds, tens, and ones.   b. Draw 12 tens 4 ones.

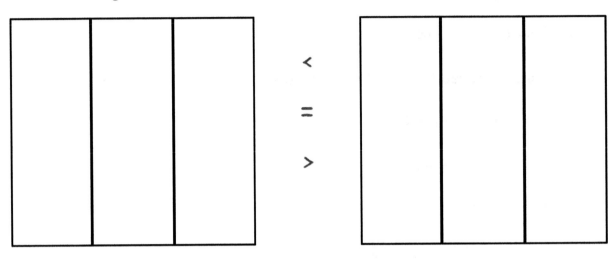

<

=

>

2. Write >, <, or =.

    a. 1 hundred 6 tens   ◯   106

    b. 74 tens   ◯   700 + 4

    c. Thirty tens   ◯   300

    d. 21 ones 3 hundreds   ◯   31 tens

Lesson 17:   Compare two three-digit numbers using <, >, and = when there are
more than 9 ones or 9 tens.

EUREKA
MATH™

Name _____   Date _____

1. Whisper count as you show the numbers with place value disks. Circle >, <, or =.

    a. Draw 13 ones and 2 hundreds.   b. Draw 12 tens and 8 ones.

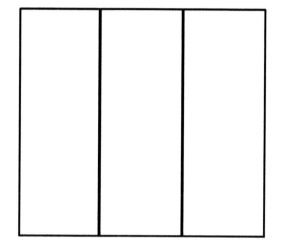

         <
         =
         >

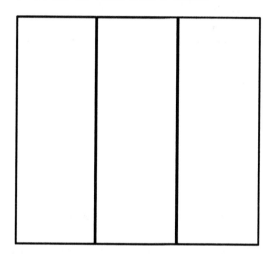

2. Write >, <, or =.

    a. 199 ◯ 10 tens   g. 400 + 2 + 50 ◯ 524

    b. 236 ◯ 23 tens 5 ones   h. 59 tens + 2 tens ◯ 610

    c. 21 tens ◯ Two hundred twenty   i. 506 ◯ 50 tens

    d. 380 ◯ 3 hundred 8 tens   j. 97 tens – 12 tens ◯ 85

    e. 20 + 4 + 500 ◯ 2 ones 45 tens   k. 67 tens + 10 tens ◯ 7 hundreds 7 ones

    f. 600 + 7 ◯ 76 tens   l. 8 hundreds 13 ones ◯ 75 tens

©2015 Great Minds. eureka-math.org
G2-M3-TE-B2-1.3.1-01.2016

Lesson 18

Objective: Order numbers in different forms. (Optional)

Suggested Lesson Structure

■ Fluency Practice   (12 minutes)
▦ Application Problem   (8 minutes)
▦ Concept Development   (30 minutes)
▦ Student Debrief   (10 minutes)

  Total Time   **(60 minutes)**

Fluency Practice (12 minutes)

- Sprint: Sums–Crossing Ten **2.OA.2**   (12 minutes)

Sprint: Sums–Crossing Ten (12 minutes)

Materials: (S) Sprint: Sums–Crossing Ten Sprint

This is the third day of the sums and differences intensive. Students remember the promise that yesterday's Spring would be repeated today, and now they see that the promise has been fulfilled.. Start the session by asking them to remember how many problems they were able to finish the day before.

  T: That is your goal. Everyone's goal is different because we are not competing with each other but with...?

  S: Ourselves!

  T: Your personal best. That is what matters. Share with your partner at least one strategy you use for practicing your sums and differences.

  S: (Share.)

  T: Here we go. Take your mark, get set, think!

Application Problem (8 minutes)

For an art project, Daniel collected 15 fewer maple leaves than oak leaves. He collected 60 oak leaves. How many maple leaves did he collect?

After guiding the students through the RDW process, let them analyze some work.

Here are some suggested questions based on the drawings to the right.

a. How does the number sentence relate to the drawing?

b. How does the first drawing relate to the second drawing?

c. What did the student who drew the place value disks do to start the problem?

d. Could the person who drew the number bonds also have started with making both the oak and maple leaves equal?

e. Can you see that equality in both pictures?

Concept Development (30 minutes)

Concrete (6 minutes)

Materials: (T) Unlabeled hundreds place value chart (Lesson 8 Template), place value disks (9 hundreds, tens and ones) (S) Unlabeled hundreds place value chart (Lesson 8 Template), place value disks (9 hundreds, tens and ones), personal white board

T: Slide the place value chart inside your personal white boards.

T: Partner A, show 2 hundreds 12 ones on your place value chart. Partner B, show 15 tens 4 ones.

T: (As students work, project your own place value chart and use place value disks to show 103.)

T: Compare numbers with your partner and me.

S: (Compare.)

T: What's the smallest, or least, number?

S: 103.

T: The greatest?

S: 212, or 2 hundreds 12 ones.

T: Write the three numbers from least to greatest on your personal white boards. Use standard form. At the signal, show your boards.

S: (Write 103, 154, 212.)

T: Good. Partner A, change to show 62 tens 4 ones. Partner B, change to show 4 ones 6 hundreds.

T: (As students work, show 642 on your place value chart.)

S: (Show.)

T: Now, compare. Write the numbers from least to greatest on your boards.

S: (Compare and show 604, 624, 642.)

NOTES ON
MULTIPLE MEANS
OF ENGAGEMENT:

As mentioned in Lesson 17, it is wise to provide visual support for struggling students. The teacher directives are coupled with the personal boards but are entirely oral. Write the directives while saying them aloud so that students see the connections and build toward the chart.

Partner A	Partner B
2 hundreds 12 ones	15 tens 4 ones
212	154
62 tens 4 ones	4 ones 6 hundreds
624	604
5 + 300 + 30	50 + 3 + 300
335	353
30 tens + 7 tens	29 tens + 8 tens
37 tens	37 tens

Lesson 18:   Order numbers in different forms. (Optional)   245

©2015 Great Minds. eureka-math.org
G2-M3-TE-B2-1.3.1-01.2016

T: Nice work. Partner A, change to show 5 + 300 + 30. Partner B, change to show 50 + 3 + 300.

T: (As students work, write *five hundred thirty-three* in word form instead of using place value disks.)

S: (Show.)

T: Compare our numbers. This time write them from greatest to least on your boards.

S: (Compare and show 533, 353, 335.)

T: You paid careful attention to the order, switching to go from greatest to least!

T: Partner A, change to show 30 tens + 7 tens. Partner B, change to show 29 tens + 8 tens.

T: (As students work, write *three hundred seventy* in word form.)

S: (Show.)

T: Compare our numbers. Write them using the symbols <, >, or = to make a number sentence.

S: (Compare and show 370 = 370 = 370.)

Pictorial (12 minutes)

Materials: (T): Pocket chart, 1 set of pre-cut <, >, = symbol cards (Lesson 15 Template 1)
(S) Personal white board

Assign students to groups by counting off as A, B, C, and D.

T: Write your letter on the back of your board so you don't forget it.

S: (Quickly write their letters.)

T: Think of a number, and draw it on your place value chart in the way that you choose.

T: Use hundreds, tens, and ones or any combination of those you'd like. Take about one minute.

S: (Think of numbers, and draw them in a variety of ways.)

T: A's, write your number in standard form below your drawing. B's, write numbers in unit form. C's, write them in word form, and D's, write them in expanded form.

Students are seated at the carpet.

T: (Collect three boards. Place the numbers side by side in the pocket chart with space between them.)

T: Work with your partner to order these three numbers from least to greatest on your personal white boards.

S: (Order the numbers on their boards.)

T: Let's read the numbers in order.

S: (Read.)

T: (Trade drawings for three new ones, and continue with two or three drawings at a time until each has been used at least once.)

NOTES ON MULTIPLE MEANS OF REPRESENTATION:

Thinking of a number can be challenging for students working below grade level. Provide some less intimidating ways to generate numbers:

- Digit cards
- Spinners
- Dice

Again, post the assignments with visual clues or examples, too.

Form	Example
A: Standard Form	24
B: Unit Form	4 ones 2 tens
C: Word Form	twenty-four
D: Expanded Form	20 + 4

Problem Set (12 minutes)

Students should do their personal best to complete the Problem Set within the allotted 12 minutes. For some classes, it may be appropriate to modify the assignment by specifying which problems they work on first. Some problems do not specify a method for solving. Students should solve these problems using the RDW approach used for Application Problems.

Instruct students to draw the values on the place value chart as directed on the Problem Set, and then order from least to greatest or greatest to least in standard form. Write <, >, or =.

Student Debrief (10 minutes)

Lesson Objective: Order numbers in different forms.

The Student Debrief is intended to invite reflection and active processing of the total lesson experience.

Invite students to review their solutions for the Problem Set. They should check work by comparing answers with a partner before going over answers as a class. Look for misconceptions or misunderstandings that can be addressed in the Debrief. Guide students in a conversation to debrief the Problem Set and process the lesson.

- T: Bring your Problem Sets to our Debrief.
- T: Work with your partner to carefully check your answers.
- S: (Work for two minutes.)
- T: Look at your drawings on your place value charts. Think about how your pictures are alike or different. Tell your partner.
- S: I drew them just like the words say. They're all different. → I used hundreds, tens, and ones in all of mine. → I drew them all differently, but then I wrote the numbers in standard form. → I decided to only use tens and ones to show each number.
- T: Look again. What about your drawings makes the numbers easy or difficult to compare?

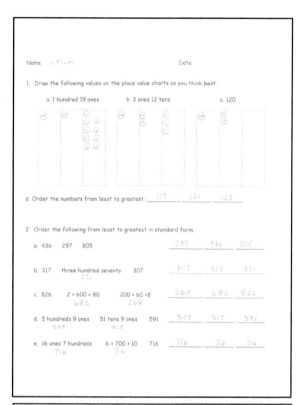

©2015 Great Minds. eureka-math.org
G2-M3-TE-B2-1.3.1-01.2016

S: It's hard to compare them when they all are in different forms. → It's also really hard when the units are mixed up.

T: How might you use what you know about comparing to help you order numbers well?

S: It helps to write all those different forms in the same way. Then, it's simple to put them in order.

T: True! Head back to your seats for your Exit Ticket.

Exit Ticket (3 minutes)

After the Student Debrief, instruct students to complete the Exit Ticket. A review of their work will help with assessing students' understanding of the concepts that were presented in today's lesson and planning more effectively for future lessons. The questions may be read aloud to the students.

©2015 Great Minds. eureka-math.org
G2-M3-TE-B2-1.3.1-01.2016

A

Number Correct: _____

Sums—Crossing Ten

1.	9 + 2 =		23.	4 + 7 =	
2.	9 + 3 =		24.	4 + 8 =	
3.	9 + 4 =		25.	5 + 6 =	
4.	9 + 7 =		26.	5 + 7 =	
5.	7 + 9 =		27.	3 + 8 =	
6.	10 + 1 =		28.	3 + 9 =	
7.	10 + 2 =		29.	2 + 9 =	
8.	10 + 3 =		30.	5 + 10 =	
9.	10 + 8 =		31.	5 + 8 =	
10.	8 + 10 =		32.	9 + 6 =	
11.	8 + 3 =		33.	6 + 9 =	
12.	8 + 4 =		34.	7 + 6 =	
13.	8 + 5 =		35.	6 + 7 =	
14.	8 + 9 =		36.	8 + 6 =	
15.	9 + 8 =		37.	6 + 8 =	
16.	7 + 4 =		38.	8 + 7 =	
17.	10 + 5 =		39.	7 + 8 =	
18.	6 + 5 =		40.	6 + 6 =	
19.	7 + 5 =		41.	7 + 7 =	
20.	9 + 5 =		42.	8 + 8 =	
21.	5 + 9 =		43.	9 + 9 =	
22.	10 + 6 =		44.	4 + 9 =	

EUREKA MATH™

Lesson 18: Order numbers in different forms. (Optional)

249

B

Number Correct: _____

Improvement: _____

Sums—Crossing Ten

1.	10 + 1 =		23.	5 + 6 =	
2.	10 + 2 =		24.	5 + 7 =	
3.	10 + 3 =		25.	4 + 7 =	
4.	10 + 9 =		26.	4 + 8 =	
5.	9 + 10 =		27.	4 + 10 =	
6.	9 + 2 =		28.	3 + 8 =	
7.	9 + 3 =		29.	3 + 9 =	
8.	9 + 4 =		30.	2 + 9 =	
9.	9 + 8 =		31.	5 + 8 =	
10.	8 + 9 =		32.	7 + 6 =	
11.	8 + 3 =		33.	6 + 7 =	
12.	8 + 4 =		34.	8 + 6 =	
13.	8 + 5 =		35.	6 + 8 =	
14.	8 + 7 =		36.	9 + 6 =	
15.	7 + 8 =		37.	6 + 9 =	
16.	7 + 4 =		38.	9 + 7 =	
17.	10 + 4 =		39.	7 + 9 =	
18.	6 + 5 =		40.	6 + 6 =	
19.	7 + 5 =		41.	7 + 7 =	
20.	9 + 5 =		42.	8 + 8 =	
21.	5 + 9 =		43.	9 + 9 =	
22.	10 + 8 =		44.	4 + 9 =	

Lesson 18: Order numbers in different forms. (Optional)

EUREKA MATH

Name _____   Date _____

1. Draw the following values on the place value charts as you think best.

   a. 1 hundred 19 ones   b. 3 ones 12 tens   c. 120

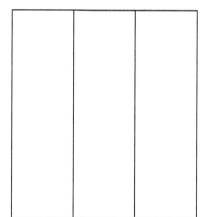 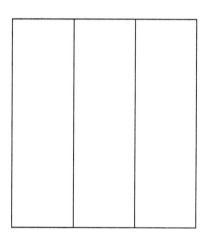 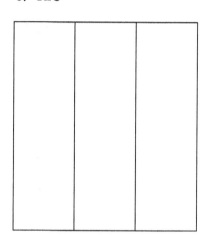

   d. Order the numbers from least to greatest: _____, _____, _____

2. Order the following from least to greatest in standard form.

   a. 436   297   805   _____, _____, _____

   b. 317   three hundred seventy   307   _____, _____, _____

   c. 826   2 + 600 + 80   200 + 60 + 8   _____, _____, _____

   d. 5 hundreds 9 ones   51 tens 9 ones   591   _____, _____, _____

   e. 16 ones 7 hundreds   6 + 700 + 10   716   _____, _____, _____

EUREKA
MATH™

Lesson 18:   Order numbers in different forms. (Optional)

251

©2015 Great Minds. eureka-math.org
G2-M3-TE-B2-1.3.1-01.2016

3. Order the following from greatest to least in standard form.

   a. 731   598   802   _____, _____, _____

   b. 82 tens   eight hundreds twelve ones   128   _____, _____, _____

   c. 30 + 3 + 300   30 tens 3 ones   300 + 30   _____, _____, _____

   d. 4 ones 1 hundred   4 tens + 10 tens   114   _____, _____, _____

   e. 19 ones 6 hundreds   196   90 + 1 + 600   _____, _____, _____

4. Write >, <, or =. Whisper the complete number sentences as you work.

   a. 700 ◯ 599 ◯ 388

   b. four hundred nine ◯ 9 + 400 ◯ 490

   c. 63 tens + 9 tens ◯ seven hundred twenty ◯ 720

   d. 12 ones 8 hundreds ◯ 2 + 80 + 100 ◯ 128

   e. 9 hundreds 3 ones ◯ 390 ◯ three hundred nine

   f. 80 tens + 2 tens ◯ 837 ◯ 3 + 70 + 800

Lesson 18:   Order numbers in different forms. (Optional)   **EUREKA MATH**

Name _____ Date _____

1. Order the following from **least to greatest** in standard form.

    a. 426   152   801   _____, _____, _____

    b. six hundred twenty   206   60 tens 2 ones   _____, _____, _____

    c. 300 + 70 + 4   3 + 700 + 40   473   _____, _____, _____

2. Order the following from **greatest to least** in standard form.

    a. 4 hundreds 12 ones   421   10 + 1 + 400   _____, _____, _____

    b. 8 ones 5 hundreds   185   5 + 10 + 800   _____, _____, _____

Lesson 18:   Order numbers in different forms. (Optional)

253

©2015 Great Minds. eureka-math.org
G2-M3-TE-B2-1.3.1-01.2016

Name _____   Date _____

1. Draw the following values on the place value charts as you think best.

    a. 241

    b. 412

    c. 124

    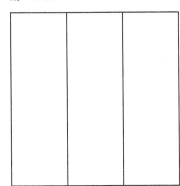
    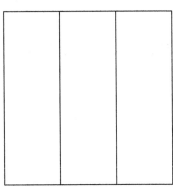
    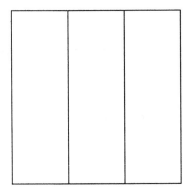

    d. Order the numbers from least to greatest: _____, _____, _____

2. Order the following from least to greatest in standard form.

    a. 537   263   912   _____, _____, _____

    b. two hundred thirty   213   20 tens 3 ones   _____, _____, _____

    c. 400 + 80 + 5   4 + 800 + 50   845   _____, _____, _____

3. Order the following from greatest to least in standard form.

    a. 11 ones 3 hundreds   311   10 + 1 + 300   _____, _____, _____

    b. 7 ones 9 hundred   79 tens + 10 tens   970   _____, _____, _____

    c. 15 ones 4 hundreds   154   50 + 1 + 400   _____, _____, _____

Lesson 18:   Order numbers in different forms. (Optional)

EUREKA MATH

Mathematics Curriculum

Topic G

Finding 1, 10, and 100 More or Less Than a Number

2.NBT.2, 2.OA.1, 2.NBT.8

Focus Standard:	2.NBT.2	Count within 1000; skip-count by 5s, 10s, and 100s
Instructional Days:	3	
Coherence -Links from:	G1–M6	Place Value, Comparison, Addition, and Subtraction to 100
-Links to:	G2–M4	Addition and Subtraction Within 200 with Word Problems to 100

The module closes with questions such as, "What number is 10 less than 402?" and "What number is 100 more than 98?" As students have been counting up and down throughout the module, these three lessons should flow nicely out of their work thus far (**2.NBT.2**). They provide a valuable transition to the addition and subtraction of the coming module, where *more* and *less* will be re-interpreted as addition and subtraction of one, ten, and a hundred (**2.NBT.8**). The language component of this segment is essential, too. Students need to be encouraged to use their words to make statements such as, "452 is 10 less than 462 and 100 less than 562." This allows for greater understanding of comparison word problems (**2.OA.1**) wherein the language of *more* and *less* is a constant presence.

A Teaching Sequence Toward Mastery of Finding 1, 10, and 100 More or Less Than a Number
Objective 1: Model and use language to tell about 1 more and 1 less, 10 more and 10 less, and 100 more and 100 less.   (Lesson 19)
Objective 2: Model 1 more and 1 less, 10 more and 10 less, and 100 more and 100 less when changing the hundreds place.   (Lesson 20)
Objective 3: Complete a pattern counting up and down.   (Lesson 21)

Lesson 19

Objective: Model and use language to tell about 1 more and 1 less, 10 more and 10 less, and 100 more and 100 less.

Suggested Lesson Structure

■ Fluency Practice   (12 minutes)
▢ Concept Development   (28 minutes)
▣ Application Problem   (10 minutes)
▨ Student Debrief   (10 minutes)

Total Time   **(60 minutes)**

Fluency Practice (12 minutes)

- Sprint: Differences **2.OA.2**   (12 minutes)

Sprint: Differences (12 minutes)

Materials: (S) Differences Sprint

T: Yesterday was our third day of practicing sums. Time to move on to differences.

T: 5 – 3 is…?

S: 2.

T: 15 – 3 is…?

S: 12.

T: 7 – 1 is…?

S: 6.

T: 17 – 1 is…?

S: 16.

T: Discuss what you see happening. How do the simple problems relate to the subtraction from the teens?

S: (Share.)

T: That is a clue to help you with today's Sprint. Take your mark, get set, think!

When closing this fluency activity, remind students that the same Sprint will be given tomorrow.

Lesson 19:   Model and use language to tell about 1 more and 1 less, 10 more and
                        10 less, and 100 more and 100 less.

Concept Development (28 minutes)

Concrete (10 minutes)

Materials: (T) Plenty of board space, sentence frames for *1 more than ___ is ___*, *10 more than ___ is ___*, and *100 more than ___ is ___* (with an analogous *less than* set) (S) Unlabeled hundreds place value chart (Lesson 8 Template), place value disks (hundreds, tens, and ones)

T: Show 110 on your place value chart.

S: (Show.)

T: Use tens disks to count by tens up to 150. (Write 150 on the board.)

S: 120, 130, 140, 150.

T: Add another tens disk.

S: (Add.)

T: 10 more than 150 is…?

S: 160.

T: (Write 160 on the board directly below 150.) Good.

T: (Post sentence frame *10 more than ___ is ___*.)
10 more than 150 is 160. Your turn.

S: 10 more than 150 is 160.

T: Add another tens disk. How many now?

S: 170.

T: (Write 170 on the board under 160.) Use the frame to say a complete sentence.

S: 10 more than 160 is 170.

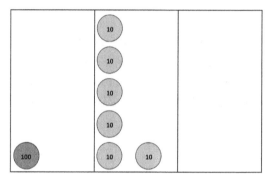

T: Look at the numbers we've counted (point to the list of 150, 160, 170). Turn and tell your partner what's the same and different about them.

S: They all have three digits. → The hundreds and ones places are the same. → The tens are changing. Every time we add a tens disk, the ten gets bigger. 5, 6, 7.

MP.8

T: I heard someone say that every time we add a tens disk, the number in the tens place grows. Use our list to predict 10 more than 170.

S: 180.

T: Using our sentence frame?

S: 10 more than 170 is 180.

T: Good. Add the tens disk to show 180.

S: (Show 180.)

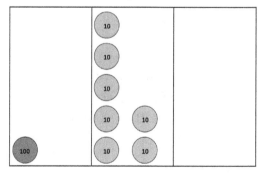

Lesson 19: Model and use language to tell about 1 more and 1 less, 10 more and 10 less, and 100 more and 100 less.

257

©2015 Great Minds. eureka-math.org
G2-M3-TE-B2-1.3.1-01.2016

T: (Write 180 under 170.) Now, count by ones to show 186. (Start another list on the board to the right of the tens, with 186 at the top.)

S: 181, 182, 183, 184, 185, 186.

T: (Post sentence frame *1 more than ___ is ___*.) Add another one disk. How many now?

S: 187.

T: Use our sentence frame to describe what you know. (Point to the *1 more than* frame.)

S: 1 more than 186 is 187.

T: (Write 187 on the board under 186.) Add another one disk.

S: 188.

T: Using our sentence frame?

S: 1 more than 187 is 188.

T: (Write 188 on the board under 187.) Look at our new list of numbers. What do you notice?

S: The ones are changing. → They're counting up by one each time we add a disk.

T: I'll label this list (150, 160, 170, 180) *10 more* since we counted by tens and this list (186, 187, 188) *1 more* because we counted by ones.

T: Talk to your partner about how our *1 more* and *10 more* lists are the same and different.

S: The hundreds are all the same. → In both lists, only 1 number changes. → When we count by tens, the tens place changes, same for the ones. → The numbers in both lists grow by 1 each time. → They look like they're growing by 1 in the tens list, but they're really growing by 10.

MP.8

T: (Label a *100 more* list to the left of *10 more*.) Let's count by hundreds. What place will change?

S: The hundreds place!

T: We have 188 now (write 188 at the top of the *100 more* list). Add a hundred disk.

S: (Show.)

T: How many now?

S: 288.

T: So… (prompt students by posting the frame *100 more than ___ is ___*).

S: 100 more than 188 is 288.

T: (Write 288 under 188 on the *100 more* list.) Were we right? Which place is changing?

S: The hundreds place!

T: Use the pattern to finish my sentence. 100 more than 288 is…?

S: 388.

T: (Write 388 under 288.) Good. Place another hundred disk to check and see.

NOTES ON
MULTIPLE MEANS
OF ENGAGEMENT:

English language learners may have a challenging time articulating how the *1 more* and *10 more* lists are the same and different. Encourage them to use their place value disks to help them explain their thinking if needed. Additionally, invite them to refer to the sentence frames posted on the board to support their responses.

Continue, but switch so that students practice counting down by hundreds, tens, and ones.

258   Lesson 19:   Model and use language to tell about 1 more and 1 less, 10 more and
                             10 less, and 100 more and 100 less.

EUREKA
MATH™

Pictorial (8 minutes)

T: With 1 more and 1 less, which place is changing?

S: The ones!

T: (Draw and write 427.) What number am I showing?

S: 427.

T: (Draw a one disk.) Use our frame to describe what happened. (Point to the *1 more* frame.)

S: 1 more than 427 is 428.

T: (Write 428 under 427.) 1 more than 428 is…?

T: (Draw a one disk.)

S: 429.

T: So, 1 less than 429 is…?

S: 428.

T: We can say, "1 less than 429 is 428." Your turn.

S: 1 less than 429 is 428.

T: (Draw a tens disk.) What place changed?

S: The tens!

T: Now, what's my number?

S: 439.

T: I'll add another ten (draw a tens disk). What's my number now?

S: 449.

T: So, 10 less than 449 is…?

S: 439.

T: We can say, "10 less than 449 is 439." Your turn.

S: 10 less than 449 is 439.

T: (Draw a hundred disk.) What's my number?

S: 549.

T: (Write 649 in standard form next to the drawing.) What unit should I put in order to have 649?

S: 1 hundred.

T: We can say, "100 more than 549 is 649." Your turn.

S: 100 more than 549 is 649.

T: (Write 650 next to 649.) What is the difference between 649 and 650?

S: A ten!

T: Let's think about that. Join in, and count with me.

S: (Chorally count.) 646, 647, 648, 649, 650.

T: So, what is the difference between 649 and 650?

S: 1.

T: Yes. We can say, "1 less than 650 is 649." Your turn.

> **NOTES ON MULTIPLE MEANS OF ENGAGEMENT:**
>
> If students have a hard time identifying which place value is changing, instruct them to circle, underline, or highlight the number or numbers that are changing. This enables them to explicitly see the change in the digits in the ones, tens, or hundreds place.

Lesson 19:   Model and use language to tell about 1 more and 1 less, 10 more and
                 10 less, and 100 more and 100 less.

259

S: 1 less than 650 is 649.

Continue, alternating practice between *more* and *less*.

Problem Set (10 minutes)

Students should do their personal best to complete the Problem Set within the allotted 10 minutes. For some classes, it may be appropriate to modify the assignment by specifying which problems they work on first. Some problems do not specify a method for solving. Students should solve these problems using the RDW approach used for Application Problems.

Instruct students to model each problem on the place value chart, complete the chart, and whisper the complete sentence.

Application Problem (10 minutes)

Mr. Palmer's second-grade class is collecting cans for recycling. Adrian collected 362 cans, Jade collected 392 cans, and Isaiah collected 562 cans.

a.   How many more cans did Isaiah collect than Adrian?

Extension: How many fewer cans did Adrian collect than Jade?

Lead students as necessary through the sequence of questions they need to internalize.

- What do you see?
- Can you draw something?
- What can you draw?
- What conclusions can you make from your drawing?

T: Use your RDW process.

T: Talk with your partner about different ways you can solve this problem using what you've learned.

S: I put 362 in my head and skip-counted by hundreds: 462, 562.

T: So, how many more cans did Isaiah collect than Adrian? Give me a complete sentence.

S: Isaiah collected 200 more cans than Adrian.

T: How can you show that your answer is correct?

S: I could draw bundles to show the numbers.

T: Would you please come up and show us, Stella?

T: Can someone show another way of proving that 562 is 200 more than 362?

S: I would draw a place value chart.

T: Please show us, Jesse.

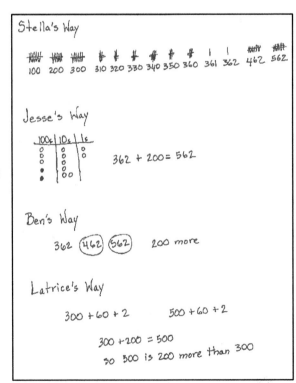

Lesson 19:   Model and use language to tell about 1 more and 1 less, 10 more and 10 less, and 100 more and 100 less.

EUREKA MATH

©2015 Great Minds. eureka-math.org
G2-M3-TE-B2-1.3.1-01.2016

T: Thank you both. Would anyone else like to share their thinking?

S: I counted on and wrote 362, 462, 562. And then, I circled how many groups of 100 I had to jump, and it was two groups, so 200.

S: I wrote it in expanded form, and it was easy to see the tens and ones were the same, but 500 is 200 more than 300.

T: I so appreciate your many ways of seeing and solving this problem! And, we all agree on the same answer, which is…?

S: Isaiah collected 200 more cans than Adrian.

T: Yes! Please complete your drawings, and add that statement to your paper.

Repeat this process with Part (b) of the question.

Student Debrief (10 minutes)

Lesson Objective: Model and use language to tell about 1 more and 1 less, 10 more and 10 less, and 100 more and 100 less.

The Student Debrief is intended to invite reflection and active processing of the total lesson experience.

Invite students to review their solutions for the Problem Set. They should check work by comparing answers with a partner before going over answers as a class. Look for misconceptions or misunderstandings that can be addressed in the Debrief. Guide students in a conversation to debrief the Problem Set and process the lesson.

T: Bring your Problem Set to our Debrief.

T: Take a couple of minutes to check over your answers with a partner.

T: Which section slowed you down? Why?

S: The fill-in-the-blank section on the Problem Set, especially (g), (h), (i), and (j). When it said *10 less,* I knew I really had to look at the tens, and when it said *100 less,* I really looked at the hundreds because those places would change.

T: Turn and tell your partner Nadia's strategy for helping herself with the fill-in-the-blank section.

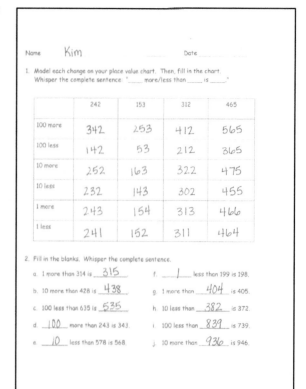

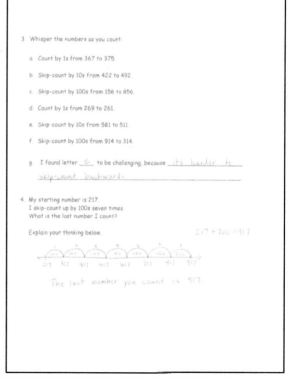

©2015 Great Minds. eureka-math.org
G2-M3-TE-B2-1.3.1-01.2016

S: Nadia paid attention to the places of numbers. → Nadia used the *10 less* and *100 less* part of the question as a clue to help her know which numbers to look at and change.

T: Let's look at Tyron and Heather's strategies for solving the last problem. (Project student work.)

T: Tyron, tell us about your strategy for solving.

S: I drew 7 lines in a row. Then, I counted by hundreds and wrote each number on a line until I filled up all the lines.

T: Thumbs up if you used the same strategy.

S: (Some show thumbs up.)

T: Now, look at Heather's strategy. Heather, can you tell us about yours?

S: I knew only the hundreds would change because we were counting by hundreds. I noticed counting by hundreds 7 times is the same as 700. I added those to the 200 in 217. I wrote 200 + 700 = 900. Then, I put 900 back together with 17 ones and got 917.

T: Good. How are these strategies the same and different?

S: They're the same because they both got the right answer. → They both only changed hundreds. → In Tyron's you can see the pattern of growing by 100. → Heather used a basic fact.

T: Pick a strategy that is different from the one you used, and try it on your paper now.

S: (Work.)

T: Good. Head back to your seats to complete your Exit Ticket.

Exit Ticket (3 minutes)

After the Student Debrief, instruct students to complete the Exit Ticket. A review of their work will help with assessing students' understanding of the concepts that were presented in today's lesson and planning more effectively for future lessons. The questions may be read aloud to the students.

©2015 Great Minds. eureka-math.org
G2-M3-TE-B2-1.3.1-01.2016

A

Number Correct: _____

Differences

1.	3 - 1 =		23.	7 - 4 =	
2.	13 - 1 =		24.	17 - 4 =	
3.	5 - 1 =		25.	7 - 5 =	
4.	15 - 1 =		26.	17 - 5 =	
5.	7 - 1 =		27.	9 - 5 =	
6.	17 - 1 =		28.	19 - 5 =	
7.	4 - 2 =		29.	7 - 6 =	
8.	14 - 2 =		30.	17 - 6 =	
9.	6 - 2 =		31.	9 - 6 =	
10.	16 - 2 =		32.	19 - 6 =	
11.	8 - 2 =		33.	8 - 7 =	
12.	18 - 2 =		34.	18 - 7 =	
13.	4 - 3 =		35.	9 - 8 =	
14.	14 - 3 =		36.	19 - 8 =	
15.	6 - 3 =		37.	7 - 3 =	
16.	16 - 3 =		38.	17 - 3 =	
17.	8 - 3 =		39.	5 - 4 =	
18.	18 - 3 =		40.	15 - 4 =	
19.	6 - 4 =		41.	8 - 5 =	
20.	16 - 4 =		42.	18 - 5 =	
21.	8 - 4 =		43.	8 - 6 =	
22.	18 - 4 =		44.	18 - 6 =	

EUREKA MATH™

Lesson 19:   Model and use language to tell about 1 more and 1 less, 10 more and 10 less, and 100 more and 100 less.

263

B

Number Correct: _____

Improvement: _____

Differences

1.	2 – 1 =	
2.	12 – 1 =	
3.	4 – 1 =	
4.	14 – 1 =	
5.	6 – 1 =	
6.	16 – 1 =	
7.	3 – 2 =	
8.	13 – 2 =	
9.	5 – 2 =	
10.	15 – 2 =	
11.	7 – 2 =	
12.	17 – 2 =	
13.	5 – 3 =	
14.	15 – 3 =	
15.	7 – 3 =	
16.	17 – 3 =	
17.	9 – 3 =	
18.	19 – 3 =	
19.	5 – 4 =	
20.	15 – 4 =	
21.	7 – 4 =	
22.	17 – 4 =	

23.	9 – 4 =	
24.	19 – 4 =	
25.	6 – 5 =	
26.	16 – 5 =	
27.	8 – 5 =	
28.	18 – 5 =	
29.	8 – 6 =	
30.	18 – 6 =	
31.	9 – 6 =	
32.	19 – 6 =	
33.	9 – 7 =	
34.	19 – 7 =	
35.	9 – 8 =	
36.	19 – 8 =	
37.	8 – 3 =	
38.	18 – 3 =	
39.	6 – 4 =	
40.	16 – 4 =	
41.	9 – 5 =	
42.	19 – 5 =	
43.	7 – 6 =	
44.	17 – 6 =	

Lesson 19:   Model and use language to tell about 1 more and 1 less, 10 more and 10 less, and 100 more and 100 less.

EUREKA MATH™

©2015 Great Minds. eureka-math.org
G2-M3-TE-B2-1.3.1-01.2016

Name _____   Date _____

1. Model each change on your place value chart. Then, fill in the chart.
   Whisper the complete sentence: "_____ more/less than _____ is _____."

	242	153	312	465
100 more				
100 less				
10 more				
10 less				
1 more				
1 less				

2. Fill in the blanks. Whisper the complete sentence.

   a. 1 more than 314 is _____.

   b. 10 more than 428 is _____.

   c. 100 less than 635 is _____.

   d. _____ more than 243 is 343.

   e. _____ less than 578 is 568.

   f. _____ less than 199 is 198.

   g. 1 more than _____ is 405.

   h. 10 less than _____ is 372.

   i. 100 less than _____ is 739.

   j. 10 more than _____ is 946.

Lesson 19:   Model and use language to tell about 1 more and 1 less, 10 more and
             10 less, and 100 more and 100 less.

265

©2015 Great Minds. eureka-math.org
G2-M3-TE-B2-1.3.1-01.2016

3. Whisper the numbers as you count:

   a. Count by 1s from 367 to 375.

   b. Skip-count by 10s from 422 to 492.

   c. Skip-count by 100s from 156 to 856.

   d. Count by 1s from 269 to 261.

   e. Skip-count by 10s from 581 to 511.

   f. Skip-count by 100s from 914 to 314.

   g. I found letter _____ to be challenging because _____

   _____.

4. My starting number is 217.

   I skip-count up by 100s seven times.

   What is the last number I count?

   Explain your thinking below.

Lesson 19: Model and use language to tell about 1 more and 1 less, 10 more and 10 less, and 100 more and 100 less.

EUREKA MATH™

©2015 Great Minds. eureka-math.org
G2-M3-TE-B2-1.3.1-01.2016

Name _____ Date _____

Fill in the blanks.

a. 10 more than 239 is _____.

b. 100 less than 524 is _____.

c. _____ more than 352 is 362.

d. _____ more than 467 is 567.

e. 1 more than _____ is 601.

f. 10 less than _____ is 241.

g. 100 less than _____ is 878.

h. 10 more than _____ is 734.

Lesson 19:   Model and use language to tell about 1 more and 1 less, 10 more and
10 less, and 100 more and 100 less.

©2015 Great Minds. eureka-math.org
G2-M3-TE-B2-1.3.1-01.2016

267

Name _____ Date _____

1. Fill in the chart. Whisper the complete sentence: "___ more/less than ___ is ___."

	146	235	357	481	672	814
100 more						
100 less						
10 more						
10 less						
1 more						
1 less						

2. Fill in the blanks. Whisper the complete sentence.

a. 1 more than 103 is _____.

b. 10 more than 378 is _____.

c. 100 less than 545 is _____.

d. _____ more than 123 is 223.

e. _____ less than 987 is 977.

f. _____ less than 422 is 421.

g. 1 more than _____ is 619.

h. 10 less than _____ is 546.

i. 100 less than _____ is 818.

j. 10 more than _____ is 974.

Lesson 19:   Model and use language to tell about 1 more and 1 less, 10 more and
              10 less, and 100 more and 100 less.

EUREKA
MATH™

Lesson 20

Objective: Model 1 more and 1 less, 10 more and 10 less, and 100 more and 100 less when changing the hundreds place.

Suggested Lesson Structure

- ■ Fluency Practice   (12 minutes)
- ■ Application Problem   (8 minutes)
- ■ Concept Development (30 minutes)
- ■ Student Debrief   (10 minutes)

   Total Time   **(60 minutes)**

Fluency Practice (12 minutes)

- ▪ Sprint: Differences **2.OA.2**   (12 minutes)

Sprint: Differences (12 minutes)

Materials: (S) Differences Sprint

T:	Today is going to be a repeat of yesterday's Sprint. Let's do some related facts practice. If I say 6 − 2, you say 16 − 2 = 14.
T:	5 − 4.
S:	15 − 4 = 11.
T:	8 − 4.
S:	18 − 4 = 14.
T:	6 − 3.
S:	16 − 3 = 13.
T:	Turn and test your partner for 30 seconds. (Pause.) Switch. (Pause.)
T:	Okay. How many of you studied last night? Are you prepared to succeed?
S:	Yes!
T:	Take your mark, get set, think!

NOTES ON MULTIPLE MEANS OF ENGAGEMENT:

It is up to the teacher to sustain a culture of personal growth and personal best in math class. This especially affects the students working below grade level and English language learners. This second day is important for them. Others may improve more than they do. Are they encouraged by their own growth? The students with the most ground to gain can truly surprise themselves if they fall in love with improving. Being proud of practicing and caring about achieving needs to become the norm.

Once again, when closing this fluency activity, inform the students that the same Sprint will be given tomorrow.

Lesson 20:   Model 1 more and 1 less, 10 more and 10 less, and 100 more and 100 less when changing the hundreds place.

©2015 Great Minds. eureka-math.org
G2-M3-TE-B2-1.3.1-01.2016

Application Problem (8 minutes)

399 jars of baby food are sitting on the shelf at the market. Some jars fall off and break. 389 jars are still on the shelf. How many jars broke?

(Lead students as necessary through the sequence of questions they need to internalize.)

- What do you see?
- Can you draw something?
- What can you draw?
- What conclusions can you make from your drawing?

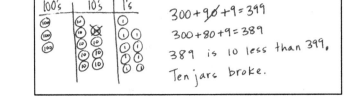

- T: Use the RDW process.
- T: Talk with your partner about different ways you can solve this problem, using what you've learned.
- S: I notice the hundreds are the same, the ones are the same, but the tens changed. So, it's 10 less.
- T: Can you draw something that will help everyone understand your thinking?
- S: I can draw a place value chart and place value disks.
- T: Please show us.
- T: Thank you, Tegan. Can someone state what Tegan said in another way?
- S: 389 is 10 less than 399.
- T: And another way?
- S: 399 is 10 more than 389.
- T: Any other thoughts?
- S: I counted on from 389 by tens: 389, 399, and my partner counted back to check: 399, 389.
- T: So, what is the answer to the question? How many jars broke?
- S: 10 jars broke.
- T: Please add that statement to your paper.

Concept Development (30 minutes)

Materials: (S) Unlabeled hundreds place value chart (Lesson 8 Template), place value disks (hundreds, tens, ones)

Concrete (10 minutes)

- T: Show 50 on your place value chart.
- S: (Show.)
- T: Use place value disks to count by ones from 50 to 59.
- S: 51, 52, 53, 54, 55, 56, 57, 58, 59.

NOTES ON MULTIPLE MEANS OF ACTION AND EXPRESSION:

The complexity of moving 10 less and changing the hundreds place together can be a big jump for some students. Therefore, use the language of tens for the following problem:

What is 10 less than 508?

T: How many tens are in 508?

S: 50 tens.

EUREKA MATH™

T: Using a complete sentence, say the number that is 1 more than 59.

S: 1 more than 59 is 60. 60 is 1 more than 59.

T: Good. Add your disk to check. Can you make a new unit?

S: (Add a disk.) Yes, a ten!

T: Trade your ones for a ten.

S: (Trade to show 6 ten disks.)

T: Use place value disks to skip-count by tens from 60 to 90.

S: 70, 80, 90.

T: Using a complete sentence, say the number that is 10 more than 90.

S: 100 is 10 more than 90. 10 more than 90 is 100.

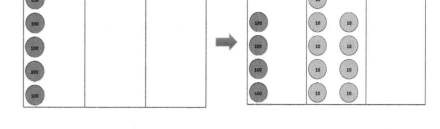

T: Add a disk to check. Can you make a new unit?

S: (Add a disk.) Yes, a hundred.

T: Make the trade.

S: (Trade to show 1 hundred disk on their place value charts.)

T: Use place value disks to skip-count by hundreds from 100 to 600.

S: 200, 300, 400, 500, 600.

T: Using a complete sentence, say the number that is 100 less than 600.

S: 500 is 100 less than 600. 100 less than 600 is 500.

T: Use your place value disks to confirm.

S: (Confirm.)

T: How can you show me ten less than 500 with your disks?

S: Trade 1 hundred for 10 tens.

T: Perfect. (Pause.) Now, can you find 10 less?

S: Yes! It's 490.

T: Show me 500 again. (Pause.) Show me 503. (Pause.)

T: How can you show me 10 less than 503?

S: The same way. Change 1 hundred for 10 tens.

T: Do you need to change the 3 ones?

S: No! Don't touch them. (Pause.)

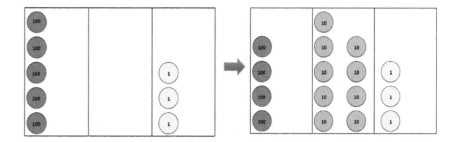

T: What is 10 less than 503?

S: 493.

©2015 Great Minds. eureka-math.org
G2-M3-TE-B2-1.3.1-01.2016

T: Give me a complete sentence.

S: 10 less than 503 is 493.

T: 10 less than 500 is…?

S: 10 less than 500 is 490.

T: 10 less than 503 is…?

S: 10 less than 503 is 493.

T: 10 more than 490 is…?

S: 10 more than 490 is 500.

T: 10 more than 493 is…?

S: 10 more than 493 is 503.

Repeat that process with a few other numbers. A suggested sequence might be 10 less than 204, 10 less than 305, 10 less than 502, and 10 less than 307. Be aware that this is setting a wonderful foundation for regrouping in subtraction and addition.

Pictorial (8 minutes)

Materials: (S) Unlabeled hundreds place value chart (Lesson 8 Template), personal white board

Begin with the place value chart inside each student's personal board.

T: Draw 130.

S: (Draw.)

T: Make it 140.

S: (Draw.)

T: Make it 150.

S: (Draw.)

T: Name my count. 1 more, 1 less, 10 more, 10 less, 100 more, or 100 less?

S: 10 more!

T: Good. Erase. Draw 715.

S: (Draw.)

T: Make it 705.

S: (Draw.)

T: Make it 695.

S: (Draw.)

T: Name my count.

S: 10 less!

Students catch on quickly. Complete another round or two, and transition into having students play with a partner while the teacher meets with a small group.

Lesson 20:   Model 1 more and 1 less, 10 more and 10 less, and 100 more and 100 less when changing the hundreds place.

©2015 Great Minds. eureka-math.org
G2-M3-TE-B2-1.3.1-01.2016

Problem Set (12 minutes)

Students should do their personal best to complete the Problem Set within the allotted 12 minutes. For some classes, it may be appropriate to modify the assignment by specifying which problems they work on first. Some problems do not specify a method for solving. Students should solve these problems using the RDW approach used for Application Problems.

Instruct students to model the problems on the place value chart, fill in the blanks, and circle all that apply. They should also whisper the complete sentence.

Student Debrief (10 minutes)

Lesson Objective: Model 1 more and 1 less, 10 more and 10 less, and 100 more and 100 less when changing the hundreds place.

The Student Debrief is intended to invite reflection and active processing of the total lesson experience.

Invite students to review their solutions for the Problem Set. They should check work by comparing answers with a partner before going over answers as a class. Look for misconceptions or misunderstandings that can be addressed in the Debrief. Guide students in a conversation to debrief the Problem Set and process the lesson.

T: Bring your Problem Set to the carpet. Skip-count down by hundreds as you transition, starting with 904.

S: 904, 804, 704, 604, 504, 404,…

T: Check your work with a partner.

S: (Compare answers.)

T: I'm hearing several of you disagree about how many jumps Jenny has to do to count to 147. Some say 7, and some say 8. Jackie, will you share your thinking?

MP.3

S: I did the difference between tens since she was counting by tens. In 77, there are 7 tens, and in 147, there are 14. I know 7 tens + 7 tens is 14 tens. That means 7 jumps.

Lesson 20: Model 1 more and 1 less, 10 more and 10 less, and 100 more and 100 less when changing the hundreds place.

273

©2015 Great Minds. eureka-math.org
G2-M3-TE-B2-1.3.1-01.2016

T: Freddy, I notice you got a different answer. Will you share your thinking?

S: I wrote the number sequence starting at 77 and finishing at 147. Then, I counted the numbers to see how many jumps. There were 8.

T: Turn and talk to your partner. Why did Freddy and Jackie get different answers?

S: Jackie did a plus problem, and Freddy counted by tens. → Jackie's right because 7 + 7 is 14, but Freddy's right, too. There are 8 numbers in his sequence. → Freddy counted Jenny's first jump! Jackie didn't. She counted on from 7: 8, 9, 10, 11, 12, 13 14. That's only 7. → Are they both right? → I think so. They just counted differently. → Jackie's answer is how many more jumps, and Freddy's answer is how many in all.

T: Many of you noticed that Freddy and Jackie both got the math right, even if they got different answers. Freddy counted how many jumps in all, and Jackie counted how many from 77. Which solution matches the way we count on? 7 or 8?

S: 7. We usually don't count the number we start with.

T: True.

Exit Ticket (3 minutes)

After the Student Debrief, instruct students to complete the Exit Ticket. A review of their work will help with assessing students' understanding of the concepts that were presented in today's lesson and planning more effectively for future lessons. The questions may be read aloud to the students.

Lesson 20:   Model 1 more and 1 less, 10 more and 10 less, and 100 more and 100
                              less when changing the hundreds place.

©2015 Great Minds. eureka-math.org
G2-M3-TE-B2-1.3.1-01.2016

A

Number Correct: _____

Differences

1.	3 – 1 =		23.	7 – 4 =	
2.	13 – 1 =		24.	17 – 4 =	
3.	5 – 1 =		25.	7 – 5 =	
4.	15 – 1 =		26.	17 – 5 =	
5.	7 – 1 =		27.	9 – 5 =	
6.	17 – 1 =		28.	19 – 5 =	
7.	4 – 2 =		29.	7 – 6 =	
8.	14 – 2 =		30.	17 – 6 =	
9.	6 – 2 =		31.	9 – 6 =	
10.	16 – 2 =		32.	19 – 6 =	
11.	8 – 2 =		33.	8 – 7 =	
12.	18 – 2 =		34.	18 – 7 =	
13.	4 – 3 =		35.	9 – 8 =	
14.	14 – 3 =		36.	19 – 8 =	
15.	6 – 3 =		37.	7 – 3 =	
16.	16 – 3 =		38.	17 – 3 =	
17.	8 – 3 =		39.	5 – 4 =	
18.	18 – 3 =		40.	15 – 4 =	
19.	6 – 4 =		41.	8 – 5 =	
20.	16 – 4 =		42.	18 – 5 =	
21.	8 – 4 =		43.	8 – 6 =	
22.	18 – 4 =		44.	18 – 6 =	

EUREKA
MATH™

Lesson 20: Model 1 more and 1 less, 10 more and 10 less, and 100 more 1 and 100 less when changing the hundreds place.

275

B

Differences

Number Correct: _____

Improvement: _____

1.	2 – 1 =	
2.	12 – 1 =	
3.	4 – 1 =	
4.	14 – 1 =	
5.	6 – 1 =	
6.	16 – 1 =	
7.	3 – 2 =	
8.	13 – 2 =	
9.	5 – 2 =	
10.	15 – 2 =	
11.	7 – 2 =	
12.	17 – 2 =	
13.	5 – 3 =	
14.	15 – 3 =	
15.	7 – 3 =	
16.	17 – 3 =	
17.	9 – 3 =	
18.	19 – 3 =	
19.	5 – 4 =	
20.	15 – 4 =	
21.	7 – 4 =	
22.	17 – 4 =	

23.	9 – 4 =	
24.	19 – 4 =	
25.	6 – 5 =	
26.	16 – 5 =	
27.	8 – 5 =	
28.	18 – 5 =	
29.	8 – 6 =	
30.	18 – 6 =	
31.	9 – 6 =	
32.	19 – 6 =	
33.	9 – 7 =	
34.	19 – 7 =	
35.	9 – 8 =	
36.	19 – 8 =	
37.	8 – 3 =	
38.	18 – 3 =	
39.	6 – 4 =	
40.	16 – 4 =	
41.	9 – 5 =	
42.	19 – 5 =	
43.	7 – 6 =	
44.	17 – 6 =	

Lesson 20: Model 1 more and 1 less, 10 more and 10 less, and 100 more and 100 less when changing the hundreds place.

EUREKA MATH

©2015 Great Minds. eureka-math.org
G2-M3-TE-B2-1.3.1-01.2016

Name _____   Date _____

1. Model each problem with a partner on your place value chart. Then, fill in the blanks, and circle all that apply. Explain your thinking.

   a. 1 more than 39 is _____.

      We made a _____.

      | one |
      | ten |
      | hundred |

   b. 10 more than 190 is _____.

      We made a _____.

      | one |
      | ten |
      | hundred |

   c. 10 more than 390 is _____.

      We made a _____.

      | one |
      | ten |
      | hundred |

   d. 1 more than 299 is _____.

      We made a _____.

      | one |
      | ten |
      | hundred |

   e. 10 more than 790 is _____.

      We made a _____.

      | one |
      | ten |
      | hundred |

2. Fill in the blanks. Whisper the complete sentence.

   a. 1 less than 120 is _____.

   b. 10 more than 296 is _____.

   c. 100 less than 229 is _____.

   d. _____ more than 598 is 608.

   e. _____ more than 839 is 840.

   f. _____ less than 938 is 838.

   g. 10 more than _____ is 306.

   h. 100 less than _____ is 894.

   i. 10 less than _____ is 895.

   j. 1 more than _____ is 1,000.

Lesson 20:   Model 1 more and 1 less, 10 more and 10 less, and 100 more and 100 less when changing the hundreds place.

277

3. Whisper the numbers as you count:

    a. Count by 1s from 106 to 115.

    b. Count by 10s from 467 to 527.

    c. Count by 100s from 342 to 942.

    d. Count by 1s from 325 to 318.

    e. Skip-count by 10s from 888 to 808.

    f. Skip-count by 100s from 805 to 5.

4. Jenny loves jumping rope.

    Each time she jumps, she skip-counts by 10s.

    She starts her first jump at 77, her favorite number.

    How many times does Jenny have to jump to get to 147?

    Explain your thinking below.

Lesson 20:   Model 1 more and 1 less, 10 more and 10 less, and 100 more and 100 less when changing the hundreds place.

©2015 Great Minds. eureka-math.org
G2-M3-TE-B2-1.3.1-01.2016

EUREKA
MATH™

Name _____ Date _____

1. Fill in the blanks, and circle the correct answer.

1 more than 209 is _____.

We made a _____.

| one |
| ten |
| hundred |

2. Fill in the blanks. Whisper the complete sentence.

a. 1 less than 150 is _____.

b. 10 more than 394 is _____.

c. _____ less than 607 is 597.

d. 10 more than _____ is 716.

e. 100 less than _____ is 894.

f. 1 more than _____ is 900.

EUREKA MATH

Lesson 20:   Model 1 more and 1 less, 10 more and 10 less, and 100 more and 100 less when changing the hundreds place.

279

©2015 Great Minds. eureka-math.org
G2-M3-TE-B2-1.3.1-01.2016

Name _____   Date _____

1. Fill in the blanks. Whisper the complete sentence.

    a. 1 less than 160 is _____.   e. _____ more than 691 is 701.

    b. 10 more than 392 is _____.   f. 10 more than _____ is 704.

    c. 100 less than 425 is _____.   g. 100 less than _____ is 986.

    d. _____ more than 549 is 550.   h. 10 less than _____ is 815.

2. Count the numbers aloud to a parent:

    a. Count by 1s from 204 to 212.   c. Skip-count by 10s from 582 to 632.

    b. Skip-count by 10s from 376 to 436.   d. Skip-count by 100s from 908 to 8.

3. Henry enjoys watching his pet frog hop.
    Each time his frog hops, Henry skip-counts backward by 100s.
    Henry starts his first count at 815.
    How many times does his frog have to jump to get to 15?

    Explain your thinking below.

Lesson 20:   Model 1 more and 1 less, 10 more and 10 less, and 100 more and 100
             less when changing the hundreds place.

EUREKA
MATH™

©2015 Great Minds. eureka-math.org
G2-M3-TE-B2-1.3.1-01.2016

Lesson 21

Objective: Complete a pattern counting up and down.

Suggested Lesson Structure

■ Fluency Practice   (12 minutes)
■ Application Problem   (8 minutes)
☐ Concept Development   (30 minutes)
■ Student Debrief   (10 minutes)

    Total Time   **(60 minutes)**

Fluency Practice (12 minutes)

▪ Sprint: Differences **2.OA.2**   (12 minutes)

Sprint: Differences (12 minutes)

Materials: (S) Differences Sprint

Lesson 21's Sprint is a review of the *take from ten* facts. This is in preparation for Module 4, in which students work toward mastery of the sums *and* differences to 20. Run a few extra copies to give to students to take home; quite a few will want to. For students struggling for fluency with these basic facts, find time if possible in the instructional day to time their improvement, or allow them to time themselves.

Application Problem (8 minutes)

Rahim is reading a really exciting book! He's on page 98. If he reads 10 pages every day, on what page will he be in 3 days?

Lead students as necessary through the sequence of questions we want them to internalize.

$$100 + 20 + 8 = 128$$

Rahim will be on page 128.

   ▪   What do you see?
   ▪   Can you draw something?
   ▪   What can you draw?
   ▪   What conclusions can you make from your drawing?
T: Use the RDW process.
T: Talk with your partner about different ways you can solve this problem using what you've learned.
T: (Invite students to share their work and explain their thinking. Then, encourage their classmates to ask them questions.)

Lesson 21:   Complete a pattern counting up and down.   **281**

©2015 Great Minds. eureka-math.org
G2-M3-TE-B2-1.3.1-01.2016

S: I drew bundles to show the number of pages he read, 98, and then I added 3 more bundles of 10 because he reads
10 pages every day.

S: I wrote 98, and then I drew 3 circles to be the 3 days and put
10 in each to show the pages he read every day. Then, I skip-counted by 10.

S: I drew a place value chart and place value disks to show 98. Then, I added a ten disk for the first day, and then a ten disk for the second day, and a ten disk for the third day because he reads 10 pages every day.

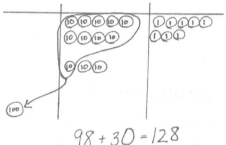

$$98 + 30 = 128$$

Rahim will be on page 128.

T: These are wonderfully clear drawings, and I like the way you explained how each piece relates to the story problem.

T: All three of these drawings help us see the pattern. Can someone explain how the numbers changed?

S: They got bigger by 10.

T: So, how were we counting?

S: We were skip-counting by 10.

T: What page will Rahim be on in 3 days?

S: Rahim will be on page 128.

T: Please add this statement to your paper.

Concept Development (30 minutes)

Concrete (10 minutes)

Materials: (S) Unlabeled hundreds place value chart (Lesson 8 Template), place value disks (hundreds, tens, and ones) per pair

T: Show 266 with place value disks.

S: (Show.)

T: Use place value disks to count out loud by ones from 266 to 272.

S: 267, 268, 269, 270, 271, 272.

T: What unit can you make?

S: A ten.

T: Go ahead and trade ones for a ten.

S: (Trade.)

NOTES ON
MULTIPLE MEANS
OF ACTION AND
EXPRESSION:

When counting back by hundreds, students may stumble when the numbers change from three-digit to two-digit numbers (e.g., 178 to 78). Have students think–pair–share beforehand on a similar problem to prepare them for the change.

©2015 Great Minds. eureka-math.org
G2-M3-TE-B2-1.3.1-01.2016

T: Use place value disks to skip-count out loud by hundreds from 272 to 772.

S: 372, 472, 572, 672, 772.

T: Say the next two numbers in our pattern.

S: 872, 972.

T: Good. Use place value disks to complete another ten. Count out loud.

S: 773, 774, 775, 776, 777, 778, 779, 780.

T: Say the next two numbers in our pattern, counting up by ones.

S: 781, 782.

T: Good. Trade your ones for a ten.

S: (Trade.)

T: Use place value disks to skip-count out loud by tens from 780 to 700.

S: 770, 760, 750, 740, 730, 720, 710, 700.

T: Say the next two numbers in our pattern.

S: 690, 680.

T: Good. Change your place value chart to show 1 more than 700.

S: (Show 701.)

T: Use place value disks to count down by tens out loud from 701 to 671.

S: 691, 681, 671.

T: (Write ____, ____, 641, 631 on the board.) Say the numbers missing from our pattern.

S: 661 and 651.

T: Yes. Use place value disks to count down by hundreds out loud from 671 to 371.

S: 571, 471, 371.

T: (Write ____, ____, 71 on the board.) Say the numbers missing from our pattern.

S: 271 and 171.

T: Nice work. Use place value disks to count out loud by ones from 371 to 375.

S: 372, 373, 374, 375.

T: (Write ____, 377, ____, ____, 380 on the board.) Say the pattern, filling in the blanks.

S: 376, 377, 378, 379, 380.

NOTES ON
MULTIPLE MEANS
OF ACTION AND
EXPRESSION:

Allow students who are challenged by the Problem Set charts to work together in pairs or in small groups. Encourage them to verbalize what they notice about the pattern and movement of the numbers before coming to the Debrief. This gives them the confidence to contribute during the discussion.

Pictorial (10 minutes)

Materials: (T) Pocket chart (S) 4 large index cards per pair

Students work as partners. Each partnership belongs to group *more* or group *less*.

T: With your partner, make a number pattern. You choose if your pattern shows counting by ones, tens, or hundreds.

T: Talk to your partner and decide now. Take 15 seconds.

S: (Partners discuss and decide.)

T: Your pattern must count down if you are in the *less* group and up if you are in the *more* group.

T: Turn and confirm with your partner: "We will count down by ____," or "We will count up by ____."

S: We will count down by tens. → We will count up by hundreds. → We will count down by ones.

T: Pick a number between 40 and 600. Partner A, write the number on a card, and hold it up.

S: (Pick a number, write it, and hold up the card.)

T: Start with that number. Use the other cards to write the rest of the numbers in your sequence.

S: (Work together.)

T: On the blank side of each card, draw the number you wrote. Take two minutes.

S: (Create their cards.)

T: Stack the cards in order with the drawings face up, and bring them to the rug with your partner.

(Students are seated at the rug.)

T: Molly and Ken, share first. Bring your cards to the pocket chart.

T: Say each number, placing one drawing at a time in the pocket. Go slowly so your friends can figure out your pattern.

T: Class, count along with Molly and Ken when you've figured out their pattern.

S: (Molly and Ken count and place, others chime in.) 236, …, 336, … (all) oh! 436, 536.

T: Name Molly and Ken's pattern.

S: 100 more!

T: Ken and Molly, can you confirm?

S: That's it!

Continue whole group or have groups share to each other and rotate. Show some patterns with numbers rather than drawings. For others, show alternating numbers, and let the class fill in blanks.

Problem Set (10 minutes)

Students should do their personal best to complete the Problem Set within the allotted 10 minutes. For some classes, it may be appropriate to modify the assignment by specifying which problems they work on first. Some problems do not specify a method for solving. Students should solve these problems using the RDW approach used for Application Problems.

Instruct students to whisper the numbers as they count, find the pattern, fill in the blanks, and complete the chart.

Student Debrief (10 minutes)

Lesson Objective: Complete a pattern counting up and down.

The Student Debrief is intended to invite reflection and active processing of the total lesson experience.

Invite students to review their solutions for the Problem Set. They should check work by comparing answers with a partner before going over answers as a class. Look for misconceptions or misunderstandings that can be addressed in the Debrief. Guide students in a conversation to debrief the Problem Set and process the lesson.

- T: Bring your Problem Set to the carpet.
  Count up by tens from 456 as you transition.
- S: 466, 476, 486, 496, 506, 516, 526, ….
- T: Take a couple of minutes, and check over your answers with your partner.
- S: (Check work.)
- T: Turn and tell your partner your reaction to Problems 3(a) and (b). What did you think?
- S: It was hard! → At first, I didn't know you had to go up and down over the white spaces to get the next number. → Yeah, the up and down ones were trickiest. → I had fun. It was like a puzzle. I used clues to fit the pieces of the puzzle together.
- T: Tim, say more about what you mean about a pattern being like a puzzle.

MP.8

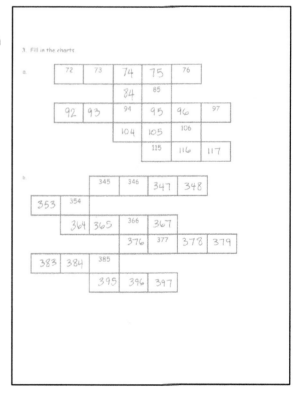

S: Well, you have to put it together in order. You have to find clues to help you figure it out.

T: What kinds of clues?

S: Like noticing if the counting is going by 1 more, 1 less, or 10 more or less, or 100 more or less.
     It makes a pattern. Once you know the pattern, it's a clue that makes things easy. The pattern just
     repeats, and you know the next number fast.

MP.8

T: Retell Tim's idea about patterns to your partner.

S: Tim said you have to look for clues about the counting. → He said you try and see if the pattern is
     going by ones, tens, or hundreds. → Tim said once you know what to count by, it's a clue that makes
     it easy to know what comes next. You just follow the pattern.

T: So, to complete number sequences like these, we look for...

S: The pattern!

T: One way that Tim did that was by noticing...

S: What the numbers are counting by!

Exit Ticket (3 minutes)

After the Student Debrief, instruct students to complete the Exit Ticket. A review of their work will help with
assessing students' understanding of the concepts that were presented in today's lesson and planning more
effectively for future lessons. The questions may be read aloud to the students.

A

Number Correct: _____

Differences

1.	10 - 5 =	
2.	10 - 0 =	
3.	10 - 1 =	
4.	10 - 9 =	
5.	10 - 8 =	
6.	10 - 2 =	
7.	10 - 3 =	
8.	10 - 7 =	
9.	10 - 6 =	
10.	10 - 4 =	
11.	10 - 8 =	
12.	10 - 3 =	
13.	10 - 6 =	
14.	10 - 9 =	
15.	10 - 0 =	
16.	10 - 5 =	
17.	10 - 7 =	
18.	10 - 2 =	
19.	10 - 4 =	
20.	10 - 1 =	
21.	11 - 1 =	
22.	11 - 2 =	

23.	11 - 3 =	
24.	10 - 9 =	
25.	11 - 9 =	
26.	10 - 5 =	
27.	11 - 5 =	
28.	10 - 7 =	
29.	11 - 7 =	
30.	10 - 8 =	
31.	11 - 8 =	
32.	10 - 6 =	
33.	11 - 6 =	
34.	10 - 4 =	
35.	11 - 4 =	
36.	10 - 9 =	
37.	12 - 9 =	
38.	10 - 5 =	
39.	12 - 5 =	
40.	10 - 7 =	
41.	12 - 7 =	
42.	10 - 8 =	
43.	12 - 8 =	
44.	15 - 9 =	

B

Number Correct: _____

Improvement: _____

Differences

1.	10 - 0 =	
2.	10 - 5 =	
3.	10 - 9 =	
4.	10 - 1 =	
5.	10 - 2 =	
6.	10 - 8 =	
7.	10 - 7 =	
8.	10 - 3 =	
9.	10 - 4 =	
10.	10 - 6 =	
11.	10 - 2 =	
12.	10 - 7 =	
13.	10 - 4 =	
14.	10 - 1 =	
15.	10 - 0 =	
16.	10 - 5 =	
17.	10 - 3 =	
18.	10 - 8 =	
19.	10 - 6 =	
20.	10 - 9 =	
21.	11 - 1 =	
22.	11 - 2 =	

23.	11 - 3 =	
24.	10 - 5 =	
25.	11 - 5 =	
26.	10 - 9 =	
27.	11 - 9 =	
28.	10 - 8 =	
29.	11 - 8 =	
30.	10 - 7 =	
31.	11 - 7 =	
32.	10 - 4 =	
33.	11 - 4 =	
34.	10 - 6 =	
35.	11 - 6 =	
36.	10 - 5 =	
37.	12 - 5 =	
38.	10 - 9 =	
39.	12 - 9 =	
40.	10 - 8 =	
41.	12 - 8 =	
42.	10 - 7 =	
43.	12 - 7 =	
44.	14 - 9 =	

Lesson 21: Complete a pattern counting up and down.

EUREKA MATH

Name _____   Date _____

1. Whisper the numbers as you count:

    a. Count by 1s from 326 to 334.

    b. Skip-count by 10s from 472 to 532.

    c. Skip-count by 10s from 930 to 860.

    d. Skip-count by 100s from 708 to 108.

2. Find the pattern. Fill in the blanks.

    a. 297, 298, _____, _____, _____, _____

    b. 143, 133, _____, _____, _____, _____

    c. 357, 457, _____, _____, _____, _____

    d. 578, 588, _____, _____, _____, _____

    e. 132, _____, 134, _____, _____, 137

    f. 409, _____, _____, 709, 809, _____

    g. 210, _____, 190, _____, _____, 160, 150

3. Fill in the charts.

a.

72	73			76	
			85		
		94			97
				106	
		115			

b.

		345	346		
	354				
		366			
			377		
		385			

    Lesson 21:   Complete a pattern counting up and down.

EUREKA MATH

Name _____ Date _____

Find the pattern. Fill in the blanks.

1. 109, _____, 111, _____, _____, 114

2. 710, _____, 690, _____, _____, 660, 650

3. 342, _____, _____, 642, 742, _____

4. 902, _____, _____, 872 , _____, 852

Name _____ Date _____

1. Find the pattern. Fill in the blanks.

a. 396, 397, _____, _____, _____, _____

b. 251, 351, _____, _____, _____, _____

c. 476, 486, _____, _____, _____, _____

d. 630, 620, _____, _____, _____, _____

e. 208, 209, _____, _____, _____, 213

f. 316, _____, _____, 616, 716, _____

g. 547, _____, 527, _____, 507, _____

h. 672, _____, 692, _____, _____

2. Fill in the chart.

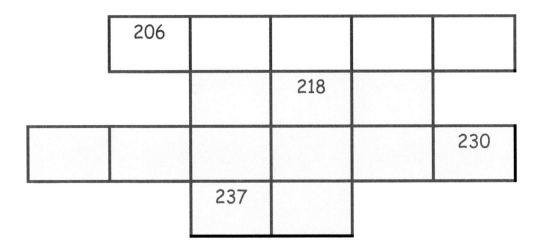

Lesson 21:   Complete a pattern counting up and down.

EUREKA
MATH™

©2015 Great Minds. eureka-math.org
G2-M3-TE-B2-1.3.1-01.2016

Name _____ Date _____

1. a. Represent 403 using place value disks.

    ┌───┐
    │ 403   │
    │   │
    │   │
    │   │
    │   │
    └───┘

    b. Write 403 in expanded form. _____

    c. Write 403 in word form. _____

2. Write each number in **standard form**.

    a. 2 hundreds 3 tens 5 ones = _____   b. 6 tens 1 hundred 8 ones = _____

    c. 600 + 4 + 30 = _____   d. 80 + 400 = _____

    e. Two hundred thirteen = _____   f. Seven hundred thirty = _____

3. Complete each statement.

    a. 10 tens = _____ hundred   b. 10 ones = _____ ten

    c. _____ tens = 1 hundred   d. 160 = _____ tens

4. Write the total amount of money shown in each group in the space below.

a.

$100	$100
$100	$100
$100	$10
$100	$10
$100	$10

b.

$10	$1
$10	$1
$10	$1
$10	$1
$10	$1

c.

$1	$100
$1	$100
$1	$100
$1	$100
$1	$100

a. _____   b. _____   c. _____

d. Write one way you can skip-count by tens and hundreds from 150 to 410.

5. Compare.

a. 456 ◯ 465

b. 10 tens ◯ 99

c. 60 + 800 ◯ Eight hundred sixteen

d. 23 tens 7 ones ◯ 237

e. 50 + 9 + 600 ◯ 9 ones 65 tens

©2015 Great Minds. eureka-math.org
G2-M3-TE-B2-1.3.1-01.2016

End-of-Module Assessment Task	Topics A–G
Standards Addressed	

Understand place value.

2.NBT.1   Understand that the three digits of a three-digit number represent amounts of hundreds, tens and ones: e.g., 706 equals 7 hundreds, 0 tens and 6 ones. Understand the following as special cases:

   a.   100 can be thought of as a bundle of ten tens—called a "hundred."

   b.   The numbers 100–900 refer to one, two, three, four, five, six, seven, eight, or nine hundreds (and 0 tens and ones).

2.NBT.2   Count within 1000: skip-count by 5s, 10s and 100s.

2.NBT.3   Read and write numbers to 1000 using base-ten numerals, number names, and expanded form.

2.NBT.4   Compare two three-digit numbers based on meanings of the hundreds, tens, and ones digits using <, =, and > symbols to record the results of comparisons.

Evaluating Student Learning Outcomes

A Progression Toward Mastery is provided to describe steps that illuminate the gradually increasing understandings that students develop on their way to proficiency. In this chart, this progress is presented from left (Step 1) to right (Step 4). The learning goal for students is to achieve Step 4 mastery. These steps are meant to help teachers and students identify and celebrate what the students CAN do now and what they need to work on next.

©2015 Great Minds. eureka-math.org
G2-M3-TE-B2-1.3.1-01.2016

A Progression Toward Mastery				
Assessment Task Item	STEP 1 Little evidence of reasoning without a correct answer. (1 Point)	STEP 2 Evidence of some reasoning without a correct answer. (2 Points)	STEP 3 Evidence of some reasoning with a correct answer or evidence of solid reasoning with an incorrect answer. (3 Points)	STEP 4 Evidence of solid reasoning with a correct answer. (4 Points)
1 **2.NBT.3**	Student is unable to solve any of the parts correctly.	Student solves one out of three parts correctly.	Student solves two out of three parts correctly.	Student correctly: <ul><li>Draws 403 in place value disks.</li><li>Writes 403 in expanded form.</li><li>Writes 403 in word form.</li></ul> Accept various representations of 403, such as 4 hundreds 3 ones or 40 tens 3 ones, etc.
2 **2.NBT.3**	Student solves one or two out of six parts correctly.	Student solves three or four out of six parts correctly.	Student solves five out of six parts correctly.	Student correctly answers: a. 235 b. 168 c. 634 d. 480 e. 213 f. 730
3 **2.NBT.1**	Student solves one out of four parts correctly.	Student solves two out of four parts correctly.	Student solves three out of four parts correctly.	Student correctly answers: a. 1 b. 1 c. 10 d. 16

Module 3:   Place Value, Counting, and Comparison of Numbers to 1,000

EUREKA MATH™

A Progression Toward Mastery				
4 **2.NBT.2**	Student solves one out of four parts correctly.	Student solves two out of four parts correctly.	Student solves three out of four parts correctly.	Student correctly: a. $730 b. $55 c. $505 d. Explains the answer using numbers, words, or pictures.
5 **2.NBT.4**	Student solves one out of five parts correctly.	Student solves two or three out of five parts correctly.	Student solves four out of five parts correctly.	Student correctly answers: a. < b. > c. > d. = e. =

Name __Freddy__   Date _____

1. a. Represent 403 using place value disks.

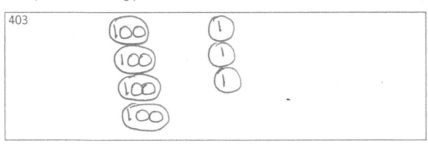

b. Write 403 in expanded form. __400 + 3__

c. Write 403 in word form. __four hundred three__

2. Write each number in **standard form**.

a. 2 hundreds 3 tens 5 ones = __235__   b. 6 tens 1 hundred 8 ones = __168__

c. 600 + 4 + 30 = __634__   d. 80 + 400 = __480__

e. Two hundred thirteen = __213__   f. Seven hundred thirty = __730__

3. Complete each statement.

a. 10 tens = __1__ hundred   b. 10 ones = __1__ ten

c. __10__ tens = 1 hundred   d. 160 = __16__ tens

Module 3:   Place Value, Counting, and Comparison of Numbers to 1,000

EUREKA
MATH™

4. Write the total amount of money shown in each group in the space below.

a.

$100	$100
$100	$100
$100	$10
$100	$10
$100	$10

b.

$10	$1
$10	$1
$10	$1
$10	$1
$10	$1

c.

$1	$100
$1	$100
$1	$100
$1	$100
$1	$100

a. __730__

b. __55__

c. __505__

d. Write one way you can skip-count by tens and hundreds from 150 to 410.

150, 160, 170, 180, 190, 200, 300, 400, 410

5. Compare.

a. 456 $<$ 465

b. 10 tens $>$ 99

c. 60 + 800 $>$ Eight hundred sixteen

d. 23 tens 7 ones $=$ 237

e. 50 + 9 + 600 $=$ 9 ones 65 tens

EUREKA
MATH™

This page intentionally left blank

Eureka Math
Grade 2
Module 3

Special thanks go to the Gordon A. Cain Center and to the Department of Mathematics at Louisiana State University for their support in the development of *Eureka Math*.

For a free *Eureka Math* Teacher Resource Pack, Parent Tip Sheets, and more please visit www.Eureka.tools

Published by the non-profit Great Minds

Printed in the U.S.A.
This book may be purchased from the publisher at eureka-math.org
10 9 8 7 6 5 4 3 2
ISBN 978-1-63255-356-0

Answer Key

GRADE 2 • MODULE 3

Place Value, Counting, and Comparison of Numbers to 1,000

Lesson 1

Problem Set

Appropriate model (ones, tens, and hundreds) for 435 drawn

Appropriate model (ones, tens, and hundreds) for 297 drawn

Appropriate model (ones, tens, and hundreds) for 673 drawn

Appropriate model (ones, tens, and hundreds) for 308 drawn

Exit Ticket

1. Lines drawn to show the following:

    10 tens = 1 hundred

    10 hundreds = 1 thousand

    10 ones = 1 ten

2. 2 hundreds circled

    9 ones boxed

3. Appropriate model drawn and labeled 627

Homework

1. 8; 8

2. 4; 40

3. 3 hundreds

    6 tens

    8 ones

4. 400; 60; 8

    468

5. Drawings will vary.

    40

EUREKA
MATH™

Lesson 2

Problem Set

1. 100 drawn, labeled, and boxed

    Units are appropriately drawn to show counting from 100 to 124; answers will vary.

2. 124 drawn, labeled, and boxed

    Units are appropriately drawn to show counting from 124 to 220; answers will vary.

3. 85 drawn, labeled, and boxed

    Units are appropriately drawn to show counting from 85 to 120; answers will vary.

4. 120 drawn, labeled, and boxed

    Units are appropriately drawn to show counting from 120 to 193; answers will vary.

Exit Ticket

1. 211 straws

    110 straws

2. Counting from 96 to 140 with ones and tens appropriately shown; pictures will vary.

3. 36, 37, 38, 39; 50, 60, 70, 80, 90; 200

Homework

1. 40, 4; 40

2. a. 14

    b. 1

    c. 140

3. Explanations will vary.

4. Counting from 68 to 130 with tens and ones appropriately shown; explanations will vary.

5. 17 bundles of 10 straws are appropriately drawn; 170

©2015 Great Minds. eureka-math.org
G2-M3-TE-B2-1.3.1-01.2016

Lesson 3

Sprint

Side A

1.	2	12.	16	23.	3	34.	11
2.	12	13.	1	24.	13	35.	1
3.	4	14.	11	25.	2	36.	11
4.	14	15.	3	26.	12	37.	4
5.	6	16.	13	27.	4	38.	14
6.	16	17.	5	28.	14	39.	1
7.	2	18.	15	29.	1	40.	11
8.	12	19.	2	30.	11	41.	3
9.	4	20.	12	31.	3	42.	13
10.	14	21.	4	32.	13	43.	2
11.	6	22.	14	33.	1	44.	12

Side B

1.	1	12.	15	23.	5	34.	12
2.	11	13.	2	24.	15	35.	1
3.	3	14.	12	25.	1	36.	11
4.	13	15.	4	26.	11	37.	5
5.	5	16.	14	27.	3	38.	15
6.	15	17.	6	28.	13	39.	2
7.	1	18.	16	29.	2	40.	12
8.	11	19.	1	30.	12	41.	4
9.	3	20.	11	31.	3	42.	14
10.	13	21.	3	32.	13	43.	1
11.	5	22.	13	33.	2	44.	11

Module 3:   Place Value, Counting, and Comparison of Numbers to 1,000

EUREKA MATH™

Problem Set

1. 90 drawn, labeled, and boxed

   Units appropriately drawn to show counting from 90 to 300; answers will vary.

2. 300 drawn, labeled, and boxed

   Units appropriately drawn to show counting from 300 to 428; answers will vary.

3. 428 drawn, labeled, and boxed

   Units appropriately drawn to show counting from 428 to 600; answers will vary.

4. 600 drawn, labeled, and boxed

   Units appropriately drawn to show counting from 600 to 1,000; answers will vary.

Exit Ticket

1. Lines drawn to show the following:

   300 to 900 matched to hundreds

   97 to 300 matched to ones and hundreds

   484 to 1,000 matched to ones, tens, and hundreds

   743 to 800 matched to ones and tens

2. Drawings will vary.

Homework

1. a. 15, 16, 17, 18, 19; 30, 40

   b. 74, 75, 76, 77, 78, 79; 90; 200; 310

   c. 66, 67, 68, 69; 80, 90

   d. 40, 50, 60, 70, 80, 90; 200, 300

2. 344

3. Ones, tens, and hundreds appropriately drawn to show counting from 668 to 900

4. a. 232 sticks

   b. Hundreds and tens appropriately drawn; 562

Lesson 4

Sprint

Side A

1.	15	12.	11	23.	15	34.	11
2.	13	13.	16	24.	13	35.	13
3.	14	14.	18	25.	18	36.	19
4.	12	15.	14	26.	17	37.	18
5.	19	16.	13	27.	16	38.	11
6.	15	17.	14	28.	15	39.	13
7.	11	18.	18	29.	15	40.	18
8.	16	19.	13	30.	13	41.	16
9.	18	20.	17	31.	12	42.	15
10.	19	21.	18	32.	19	43.	14
11.	17	22.	18	33.	16	44.	17

Side B

1.	14	12.	12	23.	15	34.	13
2.	15	13.	15	24.	16	35.	15
3.	18	14.	13	25.	13	36.	17
4.	12	15.	14	26.	12	37.	16
5.	19	16.	11	27.	18	38.	15
6.	13	17.	18	28.	18	39.	15
7.	13	18.	16	29.	12	40.	14
8.	20	19.	14	30.	16	41.	12
9.	13	20.	19	31.	19	42.	15
10.	20	21.	17	32.	18	43.	14
11.	18	22.	17	33.	16	44.	18

Module 3:   Place Value, Counting, and Comparison of Numbers to 1,000

EUREKA MATH™

Problem Set

1. Counting from 476 to 600 shown, numbers underlined where bundled to make a larger unit

2. Counting from 47 to 200 shown, numbers underlined where bundled to make a larger unit

3. Counting from 188 to 510 shown, numbers underlined where bundled to make a larger unit

4. Counting from 389 to 801 shown, numbers underlined where bundled to make a larger unit

Exit Ticket

1. c. hundred

2. 563

3. Answers will vary.

Homework

1. 257

2. 1, 0, 0

3. c. thousand

4. 5, 8, 5

5. 1, 2

6. Counting from 170 to 410 using tens and hundreds, at least 1 benchmark number circled

7. 229

Lesson 5

Problem Set

Answers will vary.

Exit Ticket

1. c. 60
2. d. 235
3. b. 168
4. 9 hundreds 5 ones

Homework

1. 700
2. a. Whole: 333; parts: 300, 30, 3; unit form: 3 hundreds 3 tens 3 ones

    b. Whole: 330; parts: 300, 30; unit form: 3 hundreds 3 tens

    c. Whole: 303; parts: 300, 3; unit form: 3 hundreds 3 ones
3. Lines drawn to show the following:

    a. 1 hundred 1 one = 101

    b. 1 ten 1 one = 11

    c. 7 tens 1 one = 71

    d. 7 hundreds 1 one = 701

    e. 1 hundred 1 ten = 110

    f. 7 hundreds 1 ten = 710

Lesson 6

Problem Set

1. 200 + 30 + 1 = 231
2. 300 + 10 + 2 = 312
3. 500 + 20 + 7 = 527
4. 700 + 50 + 2 = 752
5. 200 + 1 = 201
6. 300 + 10 = 310
7. 500 + 7 = 507
8. 700 + 50 = 750

9. 132
10. 312
11. 257
12. 572
13. 201
14. 103
15. 705
16. 507

Exit Ticket

1. Number Form
    a. 322
    b. 476
    c. 719
    d. 250
    e. 602
    f. 332

2. Expanded Form
    a. 900 + 70 + 4
    b. 400 + 30 + 5
    c. 30 + 5
    d. 300 + 10
    e. 700 + 3

Homework

1. Lines are drawn to match the following:
   a. 230
   b. 40
   c. 960
   d. 470
   e. 850
   f. 519
   g. 417
   h. 14
   i. 913
   j. 815
   k. 590
   l. 213
   m. 916

2.
   a. 244
   b. 399
   c. 125
   d. 650
   e. 403
   f. 976

3.
   a. 500 + 30 +3
   b. 300 + 50 + 5
   c. 60 + 7
   d. 400 + 60
   e. 800 + 1

EUREKA
MATH

Lesson 7

Sprint

Side A

1.	21	12.	230	23.	425	34.	715
2.	22	13.	340	24.	261	35.	705
3.	23	14.	450	25.	201	36.	347
4.	29	15.	560	26.	301	37.	307
5.	39	16.	670	27.	401	38.	532
6.	49	17.	780	28.	501	39.	502
7.	89	18.	235	29.	701	40.	502
8.	44	19.	345	30.	352	41.	602
9.	55	20.	456	31.	302	42.	642
10.	17	21.	567	32.	117	43.	713
11.	25	22.	678	33.	107	44.	738

Side B

1.	11	12.	120	23.	536	34.	815
2.	12	13.	230	24.	371	35.	805
3.	13	14.	340	25.	301	36.	237
4.	19	15.	450	26.	401	37.	207
5.	29	16.	560	27.	501	38.	642
6.	39	17.	670	28.	601	39.	602
7.	79	18.	345	29.	901	40.	602
8.	33	19.	456	30.	463	41.	603
9.	44	20.	567	31.	403	42.	643
10.	87	21.	678	32.	115	43.	815
11.	95	22.	789	33.	105	44.	729

Problem Set

Match Part 1

Lines are drawn to match the following:

a.   Answer provided

b.   374

c.   763

d.   204

e.   402

f.   743

g.   470

h.   903

i.   673

j.   213

k.   650

l.   930

m.   123

Match Part 2

a.   509

b.   434

c.   863

d.   509

e.   863

f.   509

g.   434

h.   863

i.   434

j.   863

k.   509

l.   434

Exit Ticket

1.   Three hundred forty-two

2.   a.   226

b.   803

c.   556

d.   863

3.   a.   170

b.   100 + 70

c.   1 hundred 7 tens

Homework

1.   a.   741

b.   700 + 40 + 1

c.   Seven hundred forty-one

2.   a.   560

b.   500 + 60

c.   Five hundred sixty

3.   30

4.   6

5.   221, 212, 122

Lesson 8

Problem Set

1. $100, $10, $10, $10, $1, $1, $1, $1, $1, $1; $100 + $30 + $6; whole: $136; parts: $100, $30, $6
2. $100, $100, $100, $100, $10, $10, $10, $10, $10, $1; $400 + $50 + $1;
   whole: $451; parts: $400, $50, $1
3. $100, $10, $10, $10, $10, $10, $10, $10, $10, $10; $100 + $90; whole: $190; parts: $100, $90
4. $100, $1, $1, $1, $1, $1, $1, $1, $1, $1; $100 + $9; whole: $109, parts: $100, $9
5. $100, $100, $100, $100, $10, $10, $10, $10, $10, $10; $400 + $60; whole: $460; parts: $400, $60
6. $100, $100, $100, $100, $1, $1, $1, $1, $1, $1; $400 + $6; whole: $406; parts: $400, $6
7. $100, $100, $100, $100, $100, $10, $10, $10, $10, $10; $500 + $50; whole: $550; parts: $500, $50
8. $100, $100, $100, $100, $100, $10, $10, $10, $10, $1; $500 + $40 + $1;
   whole: $541; parts: $500, $40, $1
9. $100, $100, $100, $100, $100, $100, $100, $100, $100, $1; $900 + $1; whole: $901; parts: $900, $1
10. $100, $100, $100, $100, $100, $100, $100, $100, $100, $10; $900 + $10;
    whole: $910; parts: $900, $10
11. $100, $100, $100, $100, $100, $100, $100, $100, $100, $100; $1000; no number bond
12. $10, $10, $10, $10, $10, $10, $10, $10, $10, $10; $100; no number bond

Exit Ticket

1. $356

   $300 + $50 + 6
2. $39
3. Draws two different ways to show $142; answers will vary

Homework

1. $91; $217
2. $100, $10, $10, $10, $10, $10, $10, $10, $1, $1; $100, $100, $10, $10, $1, $1, $1, $1, $1, $1
3. Bills appropriately drawn; $44

Module 3:   Place Value, Counting, and Comparison of Numbers to 1,000   315

©2015 Great Minds. eureka-math.org
G2-M3-TE-B2-1.3.1-01.2016

Lesson 9

Problem Set

1. Recordings will vary. Count shows 230.
2. Recordings will vary. Count shows 150.
3. Recordings will vary. Count shows 540.
4. Recordings will vary. Count shows 170.
5. Recordings will vary. Count shows 132.
6. Recordings will vary. Count shows 225.
7. Recordings will vary. Count shows 214.
8. Recordings will vary. Count shows 282.

Exit Ticket

1. Recordings will vary. Count shows 155.
2. Recordings will vary. Count shows 155.
3. Recordings will vary. Count shows 124.

    Number of hundreds, tens, and ones will vary.

Homework

1. a. $1,000
    b. $100
    c. $10
    d. $532
2. Answers will vary. Count shows $430.
3. Recordings will vary. Count shows $414.
4. $50

EUREKA
MATH™

Lesson 10

Sprint

Side A

1.	123	12.	730	23.	877	34.	849
2.	124	13.	703	24.	392	35.	729
3.	125	14.	33	25.	480	36.	305
4.	128	15.	733	26.	607	37.	500
5.	138	16.	940	27.	264	38.	600
6.	148	17.	904	28.	109	39.	1,000
7.	178	18.	44	29.	580	40.	482
8.	519	19.	944	30.	580	41.	365
9.	518	20.	870	31.	452	42.	232
10.	517	21.	807	32.	452	43.	767
11.	513	22.	77	33.	873	44.	555

Side B

1.	134	12.	780	23.	766	34.	863
2.	135	13.	708	24.	294	35.	927
3.	136	14.	88	25.	570	36.	304
4.	139	15.	788	26.	806	37.	600
5.	149	16.	920	27.	474	38.	700
6.	159	17.	902	28.	709	39.	1,000
7.	189	18.	22	29.	850	40.	610
8.	418	19.	922	30.	850	41.	811
9.	417	20.	760	31.	482	42.	832
10.	416	21.	706	32.	482	43.	157
11.	412	22.	66	33.	573	44.	645

©2015 Great Minds. eureka-math.org
G2-M3-TE-B2-1.3.1-01.2016

Problem Set

Drawings and explanations will vary. There are 100 $10 bills in $1,000.

Exit Ticket

Strategy different from previously explained; drawings and explanations will vary.

Homework

Drawings and explanations will vary. There are 100 $10 bills in a $1,000.

EUREKA
MATH™

©2015 Great Minds. eureka-math.org
G2-M3-TE-B2-1.3.1-01.2016

Lesson 11

Sprint

Side A

1.	3	12.	1	23.	6	34.	8
2.	3	13.	5	24.	2	35.	5
3.	2	14.	5	25.	10	36.	3
4.	1	15.	3	26.	10	37.	9
5.	5	16.	2	27.	8	38.	9
6.	5	17.	7	28.	2	39.	6
7.	4	18.	7	29.	7	40.	3
8.	1	19.	5	30.	7	41.	9
9.	9	20.	2	31.	4	42.	9
10.	9	21.	8	32.	3	43.	4
11.	8	22.	8	33.	8	44.	5

Side B

1.	4	12.	1	23.	5	34.	8
2.	4	13.	6	24.	2	35.	5
3.	3	14.	6	25.	10	36.	3
4.	1	15.	4	26.	10	37.	10
5.	6	16.	2	27.	8	38.	10
6.	6	17.	9	28.	2	39.	7
7.	5	18.	9	29.	7	40.	3
8.	1	19.	7	30.	7	41.	9
9.	10	20.	2	31.	4	42.	9
10.	10	21.	7	32.	3	43.	4
11.	9	22.	7	33.	8	44.	5

Problem Set

1. a. Twelve

        1 ten 2 ones

    b. One hundred twenty-four

        1 hundred 2 tens 4 ones

    c. One hundred four

        1 hundred 4 ones

    d. Two hundred ninety-nine

        2 hundreds 9 tens 9 ones

    e. Two hundred

        2 hundreds

2. a. Twenty-five

        2 tens 5 ones

    b. Two hundred fifty

        2 hundreds 5 tens

    c. Five hundred twenty

        5 hundreds 2 tens

    d. Five hundred two

        5 hundreds 2 ones

    e. Two hundred five

        2 hundreds 5 ones

    f. Thirty-six

        3 tens 6 ones

    g. Three hundred sixty

        3 hundreds 6 tens

    h. Six hundred thirty

        6 hundreds 3 tens

    i. Six hundred three

        6 hundreds 3 ones

    j. Three hundred six

        3 hundreds 6 ones

Exit Ticket

1. a. 36

    b. 360

2. a. 6 ones; 6 tens

    b. 3 tens; 3 hundreds

EUREKA
MATH™

Homework

1. a. Fifteen

   1 ten 5 ones

   b. One hundred fifty-two

   1 hundred 5 tens 2 ones

   c. One hundred two

   1 hundred 2 ones

   d. Two hundred ninety

   2 hundreds 9 tens

   e. Three hundred

   3 hundreds

2. a. Forty-two

   4 tens 2 ones

   b. Four hundred twenty

   4 hundreds 2 tens

   c. Three hundred twenty

   3 hundreds 2 tens

   d. Four hundred two

   4 hundreds 2 ones

   e. Four hundred forty-two

   4 hundreds 4 tens 2 ones

   f. Fifty-three

   5 tens 3 ones

   g. Five hundred thirty

   5 hundreds 3 tens

   h. Five hundred twenty

   5 hundreds 2 tens

   i. Five hundred three

   5 hundreds 3 ones

   j. Fifty-five

   5 tens 5 ones

Lesson 12

Sprint

Side A

1.	4	12.	20	23.	9	34.	18
2.	14	13.	7	24.	19	35.	9
3.	6	14.	17	25.	7	36.	19
4.	16	15.	9	26.	17	37.	7
5.	8	16.	19	27.	9	38.	17
6.	18	17.	10	28.	19	39.	5
7.	6	18.	20	29.	7	40.	15
8.	16	19.	10	30.	17	41.	8
9.	8	20.	20	31.	9	42.	18
10.	18	21.	10	32.	19	43.	8
11.	10	22.	20	33.	8	44.	18

Side B

1.	3	12.	19	23.	10	34.	19
2.	13	13.	8	24.	20	35.	9
3.	5	14.	18	25.	6	36.	19
4.	15	15.	10	26.	16	37.	8
5.	7	16.	20	27.	8	38.	18
6.	17	17.	9	28.	18	39.	6
7.	5	18.	19	29.	8	40.	16
8.	15	19.	9	30.	18	41.	9
9.	7	20.	19	31.	9	42.	19
10.	17	21.	10	32.	19	43.	7
11.	9	22.	20	33.	9	44.	17

EUREKA
MATH™

Problem Set

1.   Yes; 1 ten
2.   Yes; 1 hundred
3.   No; 2 ones
4.   Yes; 1 ten
5.   No; 8 ones
6.   Yes; 1 hundred

Exit Ticket

1.   Lines drawn to show the following:
     a.   10 ones = 1 ten
     b.   10 tens = 1 hundred
     c.   10 hundreds = 1 thousand
2.   3 hundreds, 4 tens, and 8 ones drawn on the place value chart
     a.   2
     b.   5
     c.   6

Homework

1.   No; 3 ones
2.   No; 8 ones
3.   Yes; 1 hundred
4.   No; 2 ones
5.   No; 7 ones
6.   Yes; 1 ten

Lesson 13

Sprint

Side A

1.	50	12.	96	23.	84	34.	97
2.	62	13.	60	24.	84	35.	70
3.	63	14.	76	25.	70	36.	93
4.	68	15.	77	26.	58	37.	72
5.	64	16.	73	27.	59	38.	53
6.	64	17.	78	28.	52	39.	80
7.	80	18.	78	29.	57	40.	86
8.	94	19.	90	30.	57	41.	92
9.	95	20.	81	31.	100	42.	65
10.	98	21.	82	32.	74	43.	100
11.	96	22.	87	33.	83	44.	100

Side B

1.	60	12.	86	23.	64	34.	76
2.	52	13.	70	24.	64	35.	60
3.	53	14.	96	25.	80	36.	74
4.	58	15.	97	26.	78	37.	83
5.	64	16.	93	27.	79	38.	92
6.	54	17.	98	28.	72	39.	90
7.	54	18.	98	29.	75	40.	68
8.	84	19.	50	30.	75	41.	73
9.	85	20.	61	31.	100	42.	87
10.	88	21.	62	32.	56	43.	100
11.	86	22.	67	33.	63	44.	100

EUREKA MATH™

Problem Set

1. Draws 7 ten disks, 2 one disks

    Says 7 tens 2 ones

    Says seventy-two

    Needs 8 more to change for 1 ten

    Needs 2 more to change for 1 hundred

2. Draws 4 hundred disks, 2 ten disks, 7 one disks

    Says 4 hundreds 2 tens 7 ones

    Says four hundred twenty-seven

    Needs 3 more to change for 1 ten

    Needs 7 more to change for 1 hundred

3. Draws 7 hundred-disks, 1 ten disk, 3 one disks

    Says 7 hundreds 1 ten 3 ones

    Says seven hundred thirteen

    Needs 7 more to change for 1 ten

    Needs 8 more to change for 1 hundred

4. Draws 1 hundred disk, 7 ten disks, 1 one disk

    Says 1 hundred 7 tens 1 one

    Says one hundred seventy-one

    Needs 9 more to change for 1 ten

    Needs 2 more to change for 1 hundred

5. Draws 1 hundred disk, 8 ten disks, 7 one disks

    Says 1 hundred 8 tens 7 ones

    Says one hundred eighty-seven

    Needs 3 more to change for 1 ten

    Needs 1 more to change for 1 hundred

6. Draws 7 hundred disks, 5 one disks

    Says 7 hundreds 5 ones

    Says seven hundred five

    Needs 5 more to change for 1 ten

    Needs 9 more to change for 1 hundred

Exit Ticket

1. a. Draws 5 hundred-disks, 6 ten-disks

    b. Draws 5 hundred-disks, 6 one-disks

2. Answers will vary.

Homework*

1. Draws 4 ten-disks, 3 one-disks

2. Draws 4 hundred-disks, 3 ten-disks

3. Draws 2 hundred-disks, 7 ten-disks

4. Draws 7 hundred-disks, 2 ten-disks

5. Draws 7 hundred-disks, 2 one-disks

6. Draws 9 hundred-disks, 3 ten-disks, 6 one-disks

*Answers are not included for the questions at the bottom of the homework page.

©2015 Great Minds. eureka-math.org
G2-M3-TE-B2-1.3.1-01.2016

Lesson 14

Sprint

Side A

1.	2	12.	16	23.	3	34.	11
2.	12	13.	1	24.	13	35.	1
3.	4	14.	11	25.	2	36.	11
4.	14	15.	3	26.	12	37.	4
5.	6	16.	13	27.	4	38.	14
6.	16	17.	5	28.	14	39.	1
7.	2	18.	15	29.	1	40.	11
8.	12	19.	2	30.	11	41.	3
9.	4	20.	12	31.	3	42.	13
10.	14	21.	4	32.	13	43.	2
11.	6	22.	14	33.	1	44.	12

Side B

1.	1	12.	15	23.	5	34.	12
2.	11	13.	2	24.	15	35.	1
3.	3	14.	12	25.	1	36.	11
4.	13	15.	4	26.	11	37.	5
5.	5	16.	14	27.	3	38.	15
6.	15	17.	6	28.	13	39.	2
7.	1	18.	16	29.	2	40.	12
8.	11	19.	1	30.	12	41.	4
9.	3	20.	11	31.	3	42.	14
10.	13	21.	3	32.	13	43.	1
11.	5	22.	13	33.	2	44.	11

EUREKA
MATH™

Problem Set

1. a. Draws 1 ten, 8 ones

    Draws 18 ones

    b. Draws 3 hundreds, 1 ten, 5 ones or

    2 hundreds, 11 tens, 5 ones

    Draws 3 hundreds, 15 ones

    c. Draws 2 hundreds, 6 ones or

    1 hundred 10 tens 6 ones

    Draws 20 tens, 6 ones

2. a. 0, 1, 8

    18

    b. 3, 1, 5

    3, 15

    c. 2, 0, 6

    20, 6

    d. 4, 1, 9

    41, 9

    e. 5, 7

    57

    f. 7, 48

    74, 8

    g. 9, 9

    90, 9

3. 40

Exit Ticket

1. a. Draws 2 hundreds, 4 tens, 1 one

    b. Draws 24 tens, 1 one

2. a. 0, 4, 5

    45

    b. 6, 8, 2

    6, 82

Homework

1. a. 1, 6

        16

    b. 2, 1, 7

        2, 17

    c. 3, 2, 0

        32, 0

    d. 1, 3, 9

        13, 9

    e. 4, 7, 3

        47, 3

    f. 6, 8

        68

    g. 8, 17

        81, 7

    h. 9, 21

        92, 1

2. Answers will vary.

EUREKA
MATH™

©2015 Great Minds. eureka-math.org
G2-M3-TE-B2-1.3.1-01.2016

Lesson 15

Sprint

Side A

1.	21	12.	230	23.	425	34.	715
2.	22	13.	340	24.	261	35.	705
3.	23	14.	450	25.	201	36.	347
4.	29	15.	560	26.	301	37.	307
5.	39	16.	670	27.	401	38.	532
6.	49	17.	780	28.	501	39.	502
7.	89	18.	235	29.	701	40.	502
8.	44	19.	345	30.	352	41.	602
9.	55	20.	456	31.	302	42.	642
10.	17	21.	567	32.	117	43.	713
11.	25	22.	678	33.	107	44.	738

Side B

1.	11	12.	120	23.	536	34.	815
2.	12	13.	230	24.	371	35.	805
3.	13	14.	340	25.	301	36.	237
4.	19	15.	450	26.	401	37.	207
5.	29	16.	560	27.	501	38.	642
6.	39	17.	670	28.	601	39.	602
7.	79	18.	345	29.	901	40.	602
8.	33	19.	456	30.	463	41.	603
9.	44	20.	567	31.	403	42.	643
10.	87	21.	678	32.	115	43.	815
11.	95	22.	789	33.	105	44.	729

Problem Set

1.   140; explanations will vary.

2.   16; explanations will vary.

3.   No; 5 more boxes are needed; explanations will vary.

4.   Answers will vary.

Exit Ticket

Explanations will vary.

Homework

1.   13

2.   3

3.   70

EUREKA
MATH™

©2015 Great Minds. eureka-math.org
G2-M3-TE-B2-1.3.1-01.2016

Lesson 16

Sprint

Side A

1.	10	12.	17	23.	10	34.	14
2.	11	13.	10	24.	11	35.	10
3.	12	14.	13	25.	12	36.	11
4.	18	15.	14	26.	16	37.	13
5.	10	16.	16	27.	10	38.	10
6.	11	17.	10	28.	11	39.	12
7.	12	18.	16	29.	12	40.	13
8.	17	19.	10	30.	15	41.	10
9.	10	20.	15	31.	10	42.	12
10.	13	21.	10	32.	11	43.	10
11.	14	22.	15	33.	12	44.	11

Side B

1.	10	12.	17	23.	10	34.	13
2.	11	13.	10	24.	11	35.	10
3.	12	14.	13	25.	12	36.	11
4.	16	15.	14	26.	15	37.	12
5.	10	16.	18	27.	10	38.	10
6.	11	17.	10	28.	11	39.	12
7.	12	18.	16	29.	12	40.	14
8.	17	19.	10	30.	14	41.	10
9.	10	20.	15	31.	10	42.	13
10.	13	21.	10	32.	11	43.	10
11.	14	22.	15	33.	12	44.	11

Problem Set

1. a. Draws 1 hundred, 3 tens, 2 ones
   b. Draws 3 hundreds, 1 ten, 2 ones
   c. Draws 2 hundreds, 1 ten, 3 ones
   d. 312
   e. 132
   f. 132, 213, 312

2. a. Less than
   b. Greater than
   c. Greater than
   d. Less than
   e. Less than
   f. Less than
   g. Greater than
   h. Greater than
   i. Less than
   j. Greater than

3. a. >
   b. <
   c. >
   d. <
   e. <
   f. >
   g. =
   h. >
   i. <
   j. =
   k. >

4. Charlie; 42 tens = 420 and 420 > 390

Exit Ticket

1. <
2. >
3. >
4. =
5. <
6. >
7. =
8. >

EUREKA
MATH

Homework

1. a. Draws 2 hundreds, 4 tens, 1 one

   b. Draws 4 hundreds, 1 ten, 2 ones

   c. Draws 1 hundred, 2 tens, 4 ones

   d. 124, 241, 412

2. a. Less than

   b. Less than

   c. Greater than

   d. Greater than

   e. Less than

   f. Greater than

3. a. >

   b. <

   c. <

   d. =

   e. >

   f. >

   g. =

   h. <

Lesson 17

Sprint

Side A

1.	11	12.	12	23.	11	34.	13
2.	12	13.	13	24.	12	35.	13
3.	13	14.	17	25.	11	36.	14
4.	16	15.	17	26.	12	37.	14
5.	16	16.	11	27.	11	38.	15
6.	11	17.	15	28.	12	39.	15
7.	12	18.	11	29.	11	40.	12
8.	13	19.	12	30.	15	41.	14
9.	18	20.	14	31.	13	42.	16
10.	18	21.	14	32.	15	43.	18
11.	11	22.	16	33.	15	44.	13

Side B

1.	11	12.	12	23.	11	34.	14
2.	12	13.	13	24.	12	35.	14
3.	13	14.	15	25.	11	36.	15
4.	19	15.	15	26.	12	37.	15
5.	19	16.	11	27.	14	38.	16
6.	11	17.	14	28.	11	39.	16
7.	12	18.	11	29.	12	40.	12
8.	13	19.	12	30.	11	41.	14
9.	17	20.	14	31.	13	42.	16
10.	17	21.	14	32.	13	43.	18
11.	11	22.	18	33.	13	44.	13

EUREKA
MATH™

Problem Set

1. a. Draws 2 hundreds, 1 ten, 7 ones
    b. Draws 21 tens, 7 ones
    c. Draws 1 hundred, 17 ones
    d. Draws 1 hundred, 1 ten, 7 ones
        Circles = for both problems

2. a. Greater than
    b. Equal to
    c. Less than
    d. Less than
    e. =
    f. <
    g. <
    h. >
    i. <
    j. >

3. a. <
    b. >
    c. =
    d. <
    e. <
    f. <
    g. =
    h. <
    i. >
    j. =
    k. >
    l. =
    m. <
    n. >
    o. >
    p. >

Exit Ticket

1. a. Draws 1 hundred, 4 tens, 2 ones
    b. Draws 12 tens, 4 ones
        Circles >

2. a. >
    b. >
    c. =
    d. >

Homework

1. a. Draws 2 hundreds, 13 ones

    b. Draws 12 tens, 8 ones

    Circles >

2. a. >

    b. >

    c. <

    d. =

    e. >

    f. <

    g. <

    h. =

    i. >

    j. >

    k. >

    l. >

EUREKA
MATH™

©2015 Great Minds. eureka-math.org
G2-M3-TE-B2-1.3.1-01.2016

Lesson 18

Sprint

Side A

1.	11	12.	12	23.	11	34.	13
2.	12	13.	13	24.	12	35.	13
3.	13	14.	17	25.	11	36.	14
4.	16	15.	17	26.	12	37.	14
5.	16	16.	11	27.	11	38.	15
6.	11	17.	15	28.	12	39.	15
7.	12	18.	11	29.	11	40.	12
8.	13	19.	12	30.	15	41.	14
9.	18	20.	14	31.	13	42.	16
10.	18	21.	14	32.	15	43.	18
11.	11	22.	16	33.	15	44.	13

Side B

1.	11	12.	12	23.	11	34.	14
2.	12	13.	13	24.	12	35.	14
3.	13	14.	15	25.	11	36.	15
4.	19	15.	15	26.	12	37.	15
5.	19	16.	11	27.	14	38.	16
6.	11	17.	14	28.	11	39.	16
7.	12	18.	11	29.	12	40.	12
8.	13	19.	12	30.	11	41.	14
9.	17	20.	14	31.	13	42.	16
10.	17	21.	14	32.	13	43.	18
11.	11	22.	18	33.	13	44.	13

Problem Set

1. a. Draws place value disks to show 119
   b. Draws place value disks to show 123
   c. Draws place value disks to show 120
   d. 119, 120, 123

2. a. 297, 436, 805
   b. 307, 317, 370
   c. 268, 682, 826
   d. 509, 519, 591
   e. 716, 716, 716

3. a. 802, 731, 598
   b. 820, 812, 128
   c. 333, 330, 303
   d. 140, 114, 104
   e. 691, 619, 196

4. a. >; >
   b. =; <
   c. =; =
   d. >; >
   e. >; >
   f. <; <

Exit Ticket

1. a. 152, 426, 801
   b. 206, 602, 620
   c. 374, 473, 743

2. a. 421, 412, 411
   b. 815, 508, 185

Homework

1. a. Draws place value disks to show 241
   b. Draws place value disks to show 412
   c. Draws place value disks to show 124
   d. 124, 241, 412

2. a. 263, 537, 912
   b. 203, 213, 230
   c. 485, 845, 854

3. a. 311, 311, 311
   b. 970, 907, 890
   c. 451, 415, 154

EUREKA MATH™

Lesson 19

Sprint

Side A

1.	2	12.	16	23.	3	34.	11
2.	12	13.	1	24.	13	35.	1
3.	4	14.	11	25.	2	36.	11
4.	14	15.	3	26.	12	37.	4
5.	6	16.	13	27.	4	38.	14
6.	16	17.	5	28.	14	39.	1
7.	2	18.	15	29.	1	40.	11
8.	12	19.	2	30.	11	41.	3
9.	4	20.	12	31.	3	42.	13
10.	14	21.	4	32.	13	43.	2
11.	6	22.	14	33.	1	44.	12

Side B

1.	1	12.	15	23.	5	34.	12
2.	11	13.	2	24.	15	35.	1
3.	3	14.	12	25.	1	36.	11
4.	13	15.	4	26.	11	37.	5
5.	5	16.	14	27.	3	38.	15
6.	15	17.	6	28.	13	39.	2
7.	1	18.	16	29.	2	40.	12
8.	11	19.	1	30.	12	41.	4
9.	3	20.	11	31.	3	42.	14
10.	13	21.	3	32.	13	43.	1
11.	5	22.	13	33.	2	44.	11

Problem Set

1. 342; 253; 412; 565
    142; 53; 212; 365
    252; 163; 322; 475
    232; 143; 302; 455
    243; 154; 313; 466
    241; 152; 311; 464

2. a. 315
    b. 438
    c. 535
    d. 100
    e. 10
    f. 1
    g. 404
    h. 382
    i. 839
    j. 936

3. a. 367, 368, 369, 370, 371, 372, 373, 374, 375
    b. 422, 432, 442, 452, 462, 472, 482, 492
    c. 156, 256, 356, 456, 556, 656, 756, 856
    d. 269, 268, 267, 266, 265, 264, 263, 262, 261
    e. 581, 571, 561, 551, 541, 531, 521, 511
    f. 914, 814, 714, 614, 514, 414, 314
    g. Answers will vary.

4. 917; explanations will vary.

Exit Ticket

a. 249
b. 424
c. 10
d. 100
e. 600
f. 251
g. 978
h. 724

Module 3:   Place Value, Counting, and Comparison of Numbers to 1,000

EUREKA
MATH™

Homework

1. 246; 335; 457; 581; 772; 914

   46; 135; 257; 381; 572; 714

   156; 245; 367; 491; 682; 824

   136; 225; 347; 471; 662; 804

   147; 236; 358; 482; 673; 815

   145; 234; 356; 480; 671; 813

2. a. 104

   b. 388

   c. 445

   d. 100

   e. 10

   f. 1

   g. 618

   h. 556

   i. 918

   j. 964

Lesson 20

Sprint

Side A

1.	2	12.	16	23.	3	34.	11
2.	12	13.	1	24.	13	35.	1
3.	4	14.	11	25.	2	36.	11
4.	14	15.	3	26.	12	37.	4
5.	6	16.	13	27.	4	38.	14
6.	16	17.	5	28.	14	39.	1
7.	2	18.	15	29.	1	40.	11
8.	12	19.	2	30.	11	41.	3
9.	4	20.	12	31.	3	42.	13
10.	14	21.	4	32.	13	43.	2
11.	6	22.	14	33.	1	44.	12

Side B

1.	1	12.	15	23.	5	34.	12
2.	11	13.	2	24.	15	35.	1
3.	3	14.	12	25.	1	36.	11
4.	13	15.	4	26.	11	37.	5
5.	5	16.	14	27.	3	38.	15
6.	15	17.	6	28.	13	39.	2
7.	1	18.	16	29.	2	40.	12
8.	11	19.	1	30.	12	41.	4
9.	3	20.	11	31.	3	42.	14
10.	13	21.	3	32.	13	43.	1
11.	5	22.	13	33.	2	44.	11

EUREKA
MATH™

Problem Set

1. a. 40

    Write and circle ten

    b. 200

    Write and circle hundred

    c. 400

    Write and circle hundred

    d. 300

    Write and circle ten and hundred

    e. 800

    Write and circle hundred

2. a. 119

    b. 306

    c. 129

    d. 10

    e. 1

    f. 100

    g. 296

    h. 994

    i. 905

    j. 999

3. a. 106, 107, 108, 109, 110, 111, 112, 113, 114, 115

    b. 467, 477, 487, 497, 507, 517, 527

    c. 342, 442, 542, 642, 742, 842, 942

    d. 325, 324, 323, 322, 321, 320, 319, 318

    e. 888, 878, 868, 858, 848, 838, 828, 818, 808

    f. 805, 705, 605, 505, 405, 305, 205, 105, 5

4. 7 times; explanations will vary.

Exit Ticket

1. 210

    ten

2. a. 149

    b. 404

    c. 10

    d. 706

    e. 994

    f. 899

Module 3:   Place Value, Counting, and Comparison of Numbers to 1,000

343

Homework

1. a. 159
    b. 402
    c. 325
    d. 1
    e. 10
    f. 694
    g. 1,086
    h. 825

2. a. 204, 205, 206, 207, 208, 209, 210, 211, 212
    b. 376, 386, 396, 406, 416, 426, 436
    c. 582, 592, 602, 612, 622, 632
    d. 908, 808, 708, 608, 508, 408, 308, 208, 108, 8

3. 8 times; explanations will vary.

EUREKA
MATH

Lesson 21

Sprint

Side A

1.	5	12.	7	23.	8	34.	6
2.	10	13.	4	24.	1	35.	7
3.	9	14.	1	25.	2	36.	1
4.	1	15.	10	26.	5	37.	3
5.	2	16.	5	27.	6	38.	5
6.	8	17.	3	28.	3	39.	7
7.	7	18.	8	29.	4	40.	3
8.	3	19.	6	30.	2	41.	5
9.	4	20.	9	31.	3	42.	2
10.	6	21.	10	32.	4	43.	4
11.	2	22.	9	33.	5	44.	6

Side B

1.	10	12.	3	23.	8	34.	4
2.	5	13.	6	24.	5	35.	5
3.	1	14.	9	25.	6	36.	5
4.	9	15.	10	26.	1	37.	7
5.	8	16.	5	27.	2	38.	1
6.	2	17.	7	28.	2	39.	3
7.	3	18.	2	29.	3	40.	2
8.	7	19.	4	30.	3	41.	4
9.	6	20.	1	31.	4	42.	3
10.	4	21.	10	32.	6	43.	5
11.	8	22.	9	33.	7	44.	5

Problem Set

1. a. 326, 327, 328, 329, 330, 331, 332, 333, 334
   b. 472, 482, 492, 502, 512, 522, 532
   c. 930, 920, 910, 900, 890, 880, 870, 860
   d. 708, 608, 508, 408, 308, 208, 108

2. a. 299, 300, 301, 302
   b. 123, 113, 103, 93
   c. 557, 657, 757, 857
   d. 598, 608, 618, 628
   e. 133; 135, 136
   f. 509, 609; 909
   g. 200; 180, 170

3. a. 74; 75; 84; 92; 93; 95; 96; 104; 105; 116; 117
   b. 347; 348; 353; 364; 365; 367; 376; 378; 379; 383; 384; 395; 396; 397

Exit Ticket

1. 110; 112, 113
2. 700; 680, 670
3. 442, 542; 842
4. 892, 882; 862

Homework

1. a. 398, 399, 400, 401
   b. 451, 551, 651, 751
   c. 496, 506, 516, 526
   d. 610, 600, 590, 580
   e. 210, 211, 212
   f. 416, 516; 816
   g. 537; 517; 497
   h. 682; 702, 712

2. 207; 208; 209; 210
   217; 219
   225; 226; 227; 228; 229
   238

EUREKA MATH